Philosophy and Sociology of Science

An Introduction

For Doe, with love

Philosophy and Sociology of Science

An Introduction

STEWART RICHARDS

Basil Blackwell

© Stewart Richards 1983, 1987

First published 1983
Second edition 1987
Basil Blackwell Ltd
108 Cowley Road, Oxford OX4 1JF, UK

British Library Cataloguing in Publication data

Richards, Stewart
Philosophy and sociology of science: an
introduction.—2nd ed.
1. Science—Philosophy   2. Science—Social aspects
I. Title
501   Q175

ISBN 0-631-15362-4

Library of Congress Cataloging-in-Publication Data

Richards, Stewart
Philosophy and sociology of science.
Bibliography: p.
Includes index.
1. Science—Philosophy.   2. Science—Social aspects.
I. Title.
Q175.R448 1986   501   86-26175

ISBN 0-631-15362-4 (pbk.)

Printed in Great Britain by Billing and Sons Ltd, Worcester

Contents

Part II: Interactions of Science and Society

Preface to the Second Edition

Reviewers of the first edition of this book were in broad agreement that it would be valuable in 'providing a first overview . . . of the philosophical and social relations of science'. But there was also near unanimity that it under-represented the new wave of research in sociology of science that arose in the mid and late 1970s and which, by the early 1980s, undoubtedly constituted an assertive new 'school'. I had been aware of this important development when writing the first edition but, because of the controversy it aroused, felt uncertain about how to treat it. As a result, I incorporated just two pages on this 'non-Mertonian' school rather tentatively in the section on Technology and the Sociology of Science in chapter 6.

In this second edition I have tried to give the sociology of scientific knowledge the scope that it now clearly demands. Yet I hope that this has not been at the expense of abandoning my original intention of writing a 'balanced' introduction to the fields under review; it is simply that the last decade has seen a significant shift in the cultivation of these fields which it is clearly my task to reflect.

There are also a good many other additions and amendments to the original text although I have resisted the temptation to get further embrangled in the sensitive issues of politics, race and sex, feeling, on reflection, that they are fairly treated as they stand. Indeed one critic, in what I think was intended as a compliment, suggested that because the book betrayed no political viewpoint its author must be a Social Democrat . . . . On the question of sex discrimination, I would add only that bias within science does, of course, still exist, but the evidence shows it to be relatively minor and decreasing. But as I made clear, the role of social factors in determining the entry of women into scientific professions is another matter entirely.

For the suggestion of a revised edition I am again grateful to René Olivieri at Basil Blackwell, and then to Kim Pickin for so enthusiastically taking it on. Several colleagues and students helped me in discussion, and I specially thank Antonio Gomez and Julie Lancashire for giving me their impressions of an earlier version of chapter 9.

Anyone who teaches undergraduates will know that the one sure way of generating discussion is to introduce controversy. I hope that the amendments made to this edition will make the book more profitable to discuss.

Stewart Richards
Wye College, University of London

Preface to the First Edition

This is a book about science written by a professional scientist. It is concerned, not with the facts and theories of any one discipline, but with broader questions of relevance to all who study or practise the natural and social sciences. Although intended primarily for science students, it may also be useful in providing a first overview for those beginning a detailed study of the philosophical and social relations of the sciences as subjects in their own right. The purpose of the selective bibliography is therefore to direct specialized readers to more advanced work.

However, few of the books listed will be read by those concerned with the practice of one particular science since, for them, theorizing about what it may or may not be, about how it has grown and how it relates to other disciplines or to wider human affairs, can be no more than a spare-time activity. Scientists are naturally reluctant to read books which appear to assume considerable background knowledge or which go into more detail than is compatible with an amateur interest. Moreover, they have seldom been shown that judicious reading 'around' their special subject can be highly advantageous.

Most of the books in the bibliography were written by professional historians, philosophers or sociologists with a 'professional' readership in mind. Few such professionals would write a book as general as this, for they tend to believe that to simplify complex issues is necessarily to misrepresent them. However, my own belief is that, for the professional scientist, an amateur's acquaintance with the wider dimensions of science is a great deal better than no acquaintance at all.

This conviction derives from an experience which I know to have been shared by many others. At fifteen I was ear-marked for a scientific career and advised to 'drop' history and literature. By the time I arrived at university, I had been expertly conditioned to recognize which academic subjects to pursue and which to avoid. Consequently I felt considerable resentment when actually required to take two subjects, history of science and logic and scientific method, which seemed unnecessary for one specializing in zoology. Time spent on these side issues was time not available for such essential topics as the classification of the platyhelminthes or the skull sutures of extinct reptiles. These were rigorous academic matters which I could be really proud to master.

Subsequently I learned that my own university experience had in fact differed from that of most science students; courses in the history and methodology of science were not matters of routine. But it was to be these subjects which stayed with me in later years, long after the factual information which had been stuffed into my head had disappeared. They provided a unique meeting ground for the often contrasting intellectual attitudes found in the sciences and the humanities, and they made sense of my more specialized studies in the context of wider extra-scientific interests. When I entered into scientific research and teaching, it seemed increasingly clear that a broader perspective on science must be introduced as an absolutely crucial component of the young scientist's education.

This book is based on a course that I have given for six years to a mixed class of natural and social science undergraduates at Wye College. It was begun during a period of study leave at the Unit for the History, Philosophy and Social Relations of Science in the University of Kent at Canterbury. I am very grateful to Ian Lucas, the Principal of Wye College, and his Committee for granting leave, and to Maurice Crosland, Professor of the History of Science at Canterbury, for welcoming me as a visitor in his Unit. For help on the book itself, I especially thank Maurice Crosland, his colleague Alec Dolby, René Olivieri of Basil Blackwell, and his several anonymous readers. Others who were kind enough to read and criticize parts of the manuscript, or with whom I had helpful discussions, were Christina Creek, Alex Douglas, Andrew Hill, Jeremy Naydler, Michael Preston and Crosbie Smith. Their advice and encouragement did much to improve the book, but of course I am alone responsible for its shortcomings.

Stewart Richards
Wye College, University of London

Acknowledgements

The following are sources for some of the illustrations. Permission to make use of them is gratefully acknowledged.

Figure 2: S. Hales, *Vegetable Staticks* (London, MacDonald reprint, 1969), plate 6. Figure 3: F. R. Bradbury (ed.), *Words and Numbers* (Edinburgh, University Press, 1969), opposite p. 84. Figure 5: H. Rose & S. Rose, *Science and Society* (London, Pelican, 1977), p. 5. Copyright ©, Hilary Rose and Stephen Rose, 1969; reprinted by permission of Penguin Books Ltd. Figures 6 and 7: D. de Solla Price, *Science since Babylon* (New Haven, Yale University Press, 1961), pp. 97, 116. Copyright ©, Derek de Solla Price, 1961. Figure 8: redrawn and modified from C. Freeman, *The Economics of Industrial Innovation* (London, Penguin, 1974), p. 314. Figures 9 and 10: E. Braun & D. Collingridge, *Technology and Survival* (London, Butterworths, 1977), pp. 42, 45; after D. H. Meadows, D. L. Meadows, J. Randers & W. W. Behrens, III, *The Limits to Growth: A Report for the Club of Rome's Project on the Predicament of Mankind* (a Potomac Associates book published by Universe Books, New York, 1972; graphics by Potomac Associates), Figures 35 and 46. Table 2: N. D. Ellis, 'The Occupation of Science', Technology and Society 5 (Bath, University Press, 1969), p. 40.

Physical science will not console me for the ignorance of morality in the time of affliction. But the science of ethics will always console me for the ignorance of the physical sciences.

Pascal, *Pensées*

. . . how mean a thing a mere Fact is, except as seen in the light of some comprehensive Truth.

Coleridge, *The Friend*

Introduction

Modern science is inseparable from the society that supports it. Although in Britain we were slow to organize public funding of scientific research and development, during the early part of this century the role of government increased steadily, and since the Second World War it has accelerated very rapidly indeed. To the extent that 'typical' twentieth-century science costs a good deal of public money, it is now undeniable that society gets the science it is willing and able to pay for.

As soon as we acknowledge an interaction between science and society – and abandon the romantic illusion which represents science as something 'pure' and altogether above the world of daily affairs – we are obliged to confront a whole series of questions to which there are no easy answers. Questions concerning the 'neutrality' of science quickly involve us in agonizing ethical dilemmas of immediate social and political significance. Those which have to do with its claim to be 'objective' and of special reliability as a body of knowledge soon relate to religious questions which for many people, even in a secular age, are still matters of ultimate importance; and those which may appear more mundane, relating perhaps to the applicability of scientific knowledge, are seen to have vital relevance in the decision-making processes of science policy. Matters such as these become more urgent almost daily. While few of them can be answered with certainty, what is certain is that no responsible answers at all can be given if science is treated as if it operated in a vacuum.

It has often been said that contemporary science is too dangerous to be left to the scientists. In the light of such recent developments as nuclear physics and genetic engineering, such a view seems reasonable enough. However, policy decisions about highly technical matters cannot be made *without* technical advice, so that the scientists are bound to have an important role to play. But a scientist's advice on delicate matters of public concern will inevitably reflect personal attitudes in areas beyond that of his particular professional competence. It is of little value to the politician to be told how much of some material there is in the atmosphere or the water supply; he wants to know how much of it is likely to be dangerous. On the first question – that of quantity – all scientists can agree; but on the second – that of quality – there may be sharp differences of opinion. Unfortunately there are many areas in which these two components cannot be clearly separated.

This is surely the crux of the matter. Society may be disinclined to yield to the scientists ultimate responsibility for its scientific and technical development, but it is obliged to trust their judgements on just those issues which are most controversial. Although controversy itself will never be eliminated, there would presumably be wide agreement that a scientific judgement is likely to be more informed, better balanced, even – we may hope – 'wiser', if proffered by one who has thought critically about his own science, about the nature of the scientific enterprise as a whole, and about the complex interactions of this enterprise with the society in which it has to function.

The basic contention of this book is that the quality of a scientist's education can have a lasting effect on the quality of the scientific judgements which he may later be called upon to make. By and large, the developmental experience of young scientists is still a narrow one of highly sophisticated technical training, as distinct from an expansive one of genuine education. The Robbins Report in 1963 recommended a broader education for scientists and technologists, but little has been done in most institutions of higher education to encourage a perspective beyond the parochial boundaries of one specialty. There have been few

> ... experiments in new combinations of subjects which have recognisably organic connections: technology, for instance with some social studies showing the more general implications of the technologist's profession; philosophy and mathematics with the history of science; and, for many students, some study of the past as well as the present state of the discipline they study.

This recommendation was made, not on educational grounds alone, but with future career prospects and adaptability in mind.

Soon after the Robbins Report *Contrary Imaginations* by Liam Hudson was published. This examined the predominance of the 'convergent' as distinct from the 'divergent' type of mind among young scientists. The schoolboys who featured in Hudson's study are now mature adults, yet it is doubtful whether the generation now emerging has benefited from its findings. Emphasis remains almost exclusively on 'information' and on a view of science as 'a highly complex body of fact and theory which the [student] must master before he can hope to make contributions of his own'. This emphasis may on the one hand be justifiable.

> On the other hand, science is not entirely scholastic. Above a fairly solid bedrock of accepted knowledge science is continuously and drastically revised . . ., the expectation of life for a [typical theory being] . . . somewhere in the range of fifteen years. Under the present system, therefore, the work of the [student] is dissimilar to that of the mature scientist, whose research involves intuition, imagination, and the capacity to take risks, as well as painstaking analysis. Granting this, it is possible . . . that scientific education, instead of counteracting . . . [the] natural inflexibility [of convergers], tends to reinforce and aggravate it. Scientific education, in other words, may have become dislocated from the world of research which it purports to serve.

This would be worrying enough in any situation, but if most scientists are indeed 'high-IQ convergers' the need to counter any innate rigidity and to widen intellectual horizons is even more imperative. Without some divergent experiences scientists may, as the joke has it, be expected to continue learning more and more about less and less until they know everything about nothing. They may, indeed, come to believe that science really does operate in a vacuum.

Turning the scientist's eye on to science itself is the most obvious, and arguably the most natural, way of tackling this problem. To do this, young scientists – who spend the bulk of their time focusing on technical details – need encouragement to examine the great edifice of scientific knowledge from the unfamiliar perspective of the historian, the philosopher and the sociologist. Without positive encouragement, the great majority of scientists will experience little or no discursive reasoning of the sort that is the very bread and butter of their colleagues studying the humanities. They will have no feeling for the place of science in intellectual and social history and will – except for an unrepresentative minority – finish their 'training period' firmly within the bounds of the 'Science' culture, and perhaps for ever outside that of the 'Arts'. Although it has become almost reflex to scorn the merest suggestion that these Two Cultures exist (or *still* exist – C. P. Snow gave his original lecture in 1959 and was by no means the first to identify the separation), many an 'arts person' still boasts of how impractical he is and of how very little he knows of science and technology. Many scientists still jump on any viewpoint which cannot be quantified, damning it as merely subjective if, indeed, it is not to be dismissed altogether in a smoke-screen of philistinism and threat.

Philosophy of science studied in the light of history offers the student a singular blending of arts and sciences, synthesis and analysis. The philosophical attitude is that which, more than any other, encourages an acquaintance with the intrinsic characteristics of science, both as a body of knowledge and as a method of enquiring about the world. The approach adopted here is one which emphasizes the structure of modern science, and pays relatively less attention to how this structure was built. Historical episodes are therefore used only to give concrete reality to the more abstract philosophical ideas, or to gain access to the mind of the working scientist.

By exploiting the complementary nature of these separate dimensions, it is then not difficult to see how familiarity with the logic of scientific argument concentrates the mind of the practising scientist onto the most important questions in his own work. If puzzle-solving is a part of the practice of science, then the scientist with a perspective in philosophy will be more likely to locate the correct pieces of the puzzle and be more astute at identifying which arrangements might have the qualities of a solution. By dissecting such theoretical particulars as truth and falsity, certainty and doubt, a sound foundation exists for more general decisions as to what science can, and what it probably cannot, hope to achieve.

Scientific achievement is almost invariably tied up with social goals. It is for this reason that no analysis of the scientific enterprise can be completed without the perspective of sociology. Only by examining specific features of the growth of science in society, in particular the nature of its tense relations with the great ethical and political issues of the day, can the scientist assess with detachment the necessity and desirability of pursuing one kind of science rather than another when funds are insufficient to support both. Given the burgeoning complexities of the modern world, it is only with an awareness of the broadest social dimensions of science and technology that the individual scientist may formulate a policy which can optimize the balance between man and the world in which he has to live.

Part I

Methods and
Philosophies of Science

1

The Structure of Science

The term 'Science' is now so widely applied that any definition risks offending those whose particular usage appears to be excluded. To say of something that it is 'scientific' is to encourage the view that it is altogether respectable and must be taken seriously. Thus we have scientific detergents, scientific beds, scientific humour, and scientific ways of learning to play the piano. The popular wish to be thought scientific is of course the result of the enormous power and prestige that science (whatever it may be) enjoys in our society. But in its more restricted, academic, usage the term is also highly prestigious and this was perhaps originally because science is the Latin word for knowledge; in this context, to be thought unscientific might obviously have serious implications. In any event, there is no shortage of scholarly books on *The Science of Society, The Science of Art,* and *The Science of Science.*

What all of these uses have in common is that they refer to knowledge of some aspect of 'the world'; that is, science appears to take the existence of the external world for granted. This is by no means a trivial consideration, the more so because no philosopher has ever been able to show to everyone's satisfaction that we can come to such a belief by argument alone. As soon as we look at the problem at all closely we find that some sort of intuitive or instinctive leap becomes necessary as a basis upon which we can apply our reason.

The instinctive leap that enables us to rationalize our belief in the external world is somewhat analagous to the idea or hypothesis which a scientist invents when trying to account for particular phenomena in the world. Its immediate appeal is that it has the effect of simplifying and organizing our experiences in such a way that, without it, we feel that they would not make sense. We can think of other plausible organizing principles, but the belief that there really is an external world has, for most people, the crucial advantage of being the simplest.

The full significance of this for our understanding of science will become clearer if we briefly consider why it is not irrational to doubt the existence of the world with which science deals, even though there would be no such thing as science if we were to do so. And by examining those aspects of our experience which we regard as the raw data of science, we may gain some insight into the advantages and limitations of this elusive discipline. For our purpose it is necessary to do so only in simple terms.

Of the processes that we say 'go on in our heads', we commonly, if rather vaguely, distinguish two kinds. There are those which we believe to be derived from things or events outside us in the world as a result of their effects upon our sense organs, and there are those which we believe to originate wholly within our own 'minds' and to have nothing to do with the external world. For the sake of simplicity we may call these sense perceptions, and thoughts or feelings, respectively. Although conventionally we regard our sense perceptions as giving us evidence of the external world, they actually occur *in* our minds. In this respect they are no different from our thoughts and feelings. Why, then, do we regard these two categories of internal experience differently? If both kinds actually do go on 'in our heads', how do we justify referring our sense perceptions, but not our thoughts and feelings, to a world external to our own minds?

This is where the idea of simplicity is so useful. If we argue that there is no justification for distinguishing between sense perceptions and thoughts or feelings, we shall conclude that 'the external world' is a delusion. The world, for each individual, will consist of himself (whatever that may be), his thoughts and feelings, and his sense perceptions, although the latter will give him no more than the fanciful impression of dreams. There will, of course, be no reason for believing in the existence of other people and we need explore the implications of this position no further to see that it would lead to enormous complexities and would be incompatible with any form of 'normal' life. It is in this sense that the 'instinctive leap' may be seen to have the important virtue of simplicity.

Once we have acknowledged instinctively the advantages of belief in an external world, including other people, we can use other people's testimony as a basis for separating our sense perceptions from our thoughts or feelings. Whereas the latter 'go on in our heads' for the most part independently of other people – that is, they are 'subjective' experiences accessible only to ourselves by introspection – the sense perceptions that we report prove to be very similar to those reported by others. When we have the *same* experience as others we report the *same*, or very nearly the same, sense perceptions. This is how we rationalize our instinctive belief in an external world, for if we all report closely similar sense perceptions, we feel justified in assuming that there must indeed be publicly accessible external objects 'behind' the various sense perceptions. These communal sense perceptions which, unlike our private thoughts or feelings, seem to constitute 'objective' knowledge of the world that all can confirm, are the raw data of science.

SCIENTIFIC LAWS

Science, therefore, studies those aspects of our knowledge of the external world upon which there can be universal agreement, at least in principle. It represents an attempt to reach maximum consensus. When scientists disagree about matters of science, they are generally not doing so on subjects such as the importance or reliability of their public-sense perceptions – this is typically the domain of philosophy – but on what their private thoughts or feelings suggest may be the best interpretation to be drawn. However, science would not have developed in the way it has if it were concerned only with agreement about *particular* objects and events in the world. The reason for the growth of scientific understanding is to be found in the interest of science in the relations and regularities exhibited by particular phenomena. These relations and regularities may be stated formally in scientific laws and it is a defining characteristic of such laws that they describe relations and regularities that are invariable. It is important though to realize that invariability in this context is not meant to imply certainty. While it is no doubt true that the ideal law would state truly invariable relations and regularities, in practice scientific laws are empirical in that they are derived from observational data. Even if these data were the subjects of universal agreement in the way of simple sense perceptions, we should have no way of knowing that the relations between them were truly invariable. At least, we could not know that this would be the case in the future. However, few philosophers now think that it is the purpose of science to seek absolute truth, and probably none believes that it actually finds it. Science rather proceeds *as if* the external world existed and, as a working principle, *as if* its laws were invariable.

There are two main classes of scientific laws. The simpler and the older class may be called laws of substance. These describe the invariable properties of materials and systems that occur in nature. Typical of this kind are the laws of botany or geology, for example, which state what it is to be a particular plant or rock. But laws of substance are fundamental also in physics and chemistry where they describe such things as elements and compounds. The empirical basis of such laws is well illustrated by the properties of water. At one time two 'invariable' properties were thought to be the freezing and boiling points of water, 0° and 100°C respectively; at least they were defined as such in setting up the Centigrade scale. But further investigation showed that pressure may exert a significant effect on these, so that in a precise law stating what water *is*, pressure would have to be considered as well. We can never be certain that other factors have not also been omitted.

The other class of laws, especially characteristic of the physical sciences, may be called laws of function. These describe the invariable relations that exist between the properties of materials and systems. Traditionally, they have often been said to concern the relation of cause and effect. This is one which predicts that *if* one event occurs *then* another event will invariably follow. The events are thus related in temporal sequence. A simple example would be Arrhenius's law which says (simply) that the velocity of a chemical reaction is a *function* of temperature; this allows us to predict that if the temperature is raised then the reaction rate will speed up. However, many laws of function really involve not so much the idea of causal sequence through time, as that of specific numerical relations. Thus Ohm's law says (again, simply) that in an electrical circuit of constant resistance, the current is *proportional* to the voltage. While either of these parameters may be made the cause of a change in the other (which in that circumstance will be the effect), what does not change regardless of which is cause and which effect, is the numerical relation between them. It is this that the law describes and from this that predictions may be made.

In the life and social sciences, laws concerning numerical relations seldom allow the prediction of results with a precision comparable to Ohm's law. This is not because the laws themselves are less invariable than those of physics but because they refer to events, the causes of which science does not yet understand. The events are the result, we say, of 'chance'. In these circumstances the laws have little or no ability to predict the outcome of individual events (as, for instance, the hereditary characteristics of a particular organism), although they can make accurate predictions of the statistical probabilities of a long series of events (such as the distribution ratios of characteristics in a population).

SCIENTIFIC THEORIES

While there is, or can be, something close to universal agreement among individuals concerning scientific laws, this is seldom the case with scientific theories or hypotheses. (For our purposes a hypothesis is merely the tentative expression of a theory). The reason for the difference is that laws state invariable relations and regularities about the world which can be confirmed empirically by any observer, whereas theories 'underlie' laws (typically as 'explanations') and generally are not so accessible to direct test by sense perception. Theories thus leave room for personal opinion.

The whole quesiton of the status of laws and theories is one of active debate in the philosophy of science, but for introductory purposes one of the best ways of gaining some understanding is to examine how they relate together. If laws ideally predict the outcome of new events (new in the sense of not previously known), theories ideally predict the occurrence of new laws. And a theory that can predict laws as yet unknown will also be able to account for the old laws; that is, both types of law, the known and the unknown, may be deduced as consequences of the theory. To say of a theory that it explains a law (or set of laws) is typically to interpret the more complex in terms of the more simple or, to put it another way, to refer the more particular to the more general. Boyle's law, which expressed the invariable relation that holds between the pressure and volume of a gas, allows an infinite number of new events to be predicted as a result of varying one or other parameter. When we say of this law that it was 'discovered' by Boyle, we seem to imply that it was waiting 'out there' in the world all the time.

Yet the law itself gives us no insight whatever into how it is that the inverse relation of pressure and volume holds. The modern *explanation* of Boyle's law was to come in the kinetic theory of gases, which construed the increasing pressure exerted by a gas as its volume is reduced, in terms of the greater frequency of collisions of its constituent molecules with the walls of the enclosing vessel. Thus the theory gives us a general picture, or model, of the gas which helps us to understand the properties stated in the particular law. It further allows us to predict the effect of raising the temperature of the gas at constant volume (a rise in pressure, as stated in Gay-Lussac's law) in terms of the increased velocity of the molecules. We tend to say of the theory that it was 'invented' (by Maxwell and others), which gives expression to our feeling that, in contrast to a law, a scientific theory is somehow a human artifact, something that is perhaps not quite real.

Although the last point gives rise to a serious and important question, it is one that need detain us here only briefly. There are two extreme views of the status of scientific theories. One is that the concepts employed in theories are typically fictions, mere instruments which support the imagination and guide predictions; as a result theories are seen as nothing but compact re-formulations of laws. The opposite view is that theoretical entities do refer to real things. According to this position atoms, or even 'elementary particles', could actually be shown to exist as concrete objects if only we had the techniques for doing so; in this respect they resemble entities such as bacteria and viruses in the germ theory of disease which could not be 'seen' when first postulated, but which we now can observe with the light or electron microscope. For some advocates of this view, there is thus considerable conviction to be found in the 'tracks' of elementary particles which are the raw data of high-energy physics (see chapter 5).

For present purposes it is perhaps most appropriate to see theories as models or analogues of 'real' systems in the external world. Theories therefore resemble certain features of the real systems but not all, and in many cases we do not, and probably cannot, know which features these are. (Of course, where there is a blatant lack of correspondence between the theory and external world, the theory would have to be abandoned.) This is a compromise position somewhere between the two extremes, which recognizes that theories are fallible constructs of the human mind. If it is to promote understanding the analogy used in the theory will usually be drawn with a system whose laws of operation are familiar; with a mechanical arrangement such as billiard balls in the case of Dalton's atomic theory, with the artificial selection of characteristics by animal breeders in the case of Darwin's theory of evolution by natural selection, or with wholly abstract mathematical systems in theoretical physics. Although we do not know how close the analogy between the theory and the real thing is and therefore must not think of the one as if it *were* the other, we recognize that the theory is incomplete but flexible, and can always be improved by new evidence. We accept that it is in the nature of scientific knowledge to be incomplete, but at the same time we assert that it is in our nature to seek some semblance of intelligibility in the world. At very least, the value of theories is that they guide us in this search.

The purpose of this short chapter has been merely to introduce the idea that science consists of different sorts of knowledge – sense perceptions concerning particular phenomena, the relations and reg-ularities exhibited by phenomena in general, and the organizing patterns that underlie these structural and functional relations. We have not examined the questions of whether these different kinds of knowledge may, or should, be arranged in some hierarchical order, nor that of how laws and theories have actually arisen in the course of scientific development. These are contentious issues to which a great deal of intellectual energy may be directed. Unfortunately, when they are tackled head-on and treated as central problems in the philosophy of science, they tend to loom larger than their special significance warrants. This has the effect of rendering obscure and complex what is, in essence, clear and simple, a situation beloved of some philosophers but one that we shall take careful steps to avoid. Our method will be to treat these problems naturally as they arise. In the course of the next four chapters they will be seen to present no special difficulties when considered in the context of the logic, general methodology and peculiar character of the scientific enterprise.

2

Scientific Argument: The Role of Logic

The invention of logic is usually credited to Aristotle. There seems to be no doubt that the enormous Greek interest in mathematics, particularly geometry, provided ideal conditions for the establishment of the so-called science of the laws of thought. Geometry itself drew upon knowledge obtained in the study of accurate measurements and the computation of areas that had earlier been developed by the Egyptians. The Greeks' contribution was to transform the somewhat static ideas about the relations that held between lines and angles into a compelling system which derived irrefutable conclusions from a number of initial definitions or axioms. In establishing the roots of what we now know as logic, Aristotle sought to invent a system which could be applied to spoken and written discourse with the same rigorous precision as he saw in geometry.

Logic may be defined more satisfactorily as the investigation of the principles of correct reasoning. It is concerned with the analysis of arguments and with the clarity of their expression, trying, as it does, to find rules which can justify our intuition as to what is valid. Traditionally logic has been taught as a branch of philosophy, although its value is by no means restricted to academic exercises, for arguments may arise in any area of daily life which can be resolved only by careful attention to the nature of the conflicting claims that are being made. If logic is to have any role in resolving arguments, such claims generally have to take the form of statements or propositions supported by evidence. They would then be seen as conclusions, and it is the purpose of logic to determine, not the quality of the evidence (this is typically a matter of 'fact'), nor indeed the quality of the conclusion, but rather the quality of the relation that exists between a conclusion and its evidence. By this we mean that logic evaluates *arguments*, and it is because arguments are an indispensable ingredient of science that scientists cannot do without logic. Rightly or wrongly, the high status accorded to scientific knowledge is the result of the widespread belief that it rests on sound logical foundations.

In this chapter, we shall deal with logic at an elementary level by selecting certain aspects which can serve as simple models for a highly complex and technical subject, and which can also have relevance to scientific practice.

DEDUCTION

In Aristotle's own system all arguments were broken down into three essential propositions, the first and the second being the premises which provided the evidence, and the third being the conclusion drawn from them. The resulting form of argument is known as a syllogism and this still serves as an elementary model for deduction, the most rigorous form of reasoning known. (Although the propositional calculus is now widely used as the vehicle for logical arguments, old-fashioned syllogisms retain the crucial advantage of universal intelligibility; they do not require familiarity with a symbolic notation which, for better or worse, many readers of this book would find intimidating, and therefore unhelpful.) The classic example of the syllogism is:

All men are mortal.
Socrates is a man.
Therefore Socrates is mortal.

This may seem simple enough, but there are many forms of syllogism which are less obvious, and in all of them it is the task of logic to determine whether the evidence presented in the premise(s) does indeed lead to the conclusion.

In deductive logic it is essential to make the distinction between the concepts of truth and falsity on the one hand and those of validity and invalidity on the other. The concepts of truth and falsity apply only to the constituents of an argument, the premise or premises, and the conclusion. The concepts of validity and invalidity apply only to the relation between the constituents, the argument or the inference itself. Thus we may reject the claim of a conclusion to be true either because we know a premise from which it is derived to be false (which generally has little to do with logic), or because we find something unsound (invalid) about the way in which the two relate.

Deductive argument involves the idea of conclusive evidence or proof. As a result, there is a compulsion in the reasoning which leaves no room for partial validity or invalidity because the premises contain, at least by implication, all of the information contained in the conclusion. Thus the conclusion *must* be true if the premises are true and the argument valid; *if* all men are indeed mortal, and *if* Socrates is indeed a man, then it necessarily follows that he too must be mortal. Conversely, if the argument is valid it is impossible for the premises to be true and the conclusion false; we could thus conclude that Socrates is *immortal* only if we failed altogether to comprehend the meaning of the two premises.

In this most famous of all syllogisms it is clear that the actual terms 'man (or men)', 'Socrates', and 'mortal', are themselves of no import-ance to the argument. The same logical compulsion would still operate if these terms were replaced by 'woman (or women)', 'Cleopatra', and 'illogical', no matter how strongly we may reject the first premise as patently false. The validity of the argument would, furthermore, be unimpaired if we replaced the terms by letters, say A, B, and C. We could then write:

All Bs are Cs.
A is a B.
Therefore A is a C.

The really vital terms are therefore the ones which we have not changed (e.g. 'all'), and we can say that what these various arguments have in common is their 'form'. Since the arguments are valid, we say that the form is valid. Or conversely, any argument with a valid form is a valid argument. We therefore know that no deductive argument having a valid form can have true premises and a false conclusion.

This point is especially important in science because it is not uncommon to be able to show, as a matter of fact, whether the constituent parts of an argument are indeed true or false. A simple example might be:

All vipers are reptiles.
All snakes are reptiles.
Therefore all snakes are vipers.

The form of this could be written:

All As are Cs.
All Bs are Cs.
Therefore all Bs are As.

Because we do happen to know that the premises are true and the conclusion false, we also know with certainty that the argument is invalid. But in all other combinations of truth values we cannot be sure. Note, incidentally, that we have merely to show that the premises *could* be true when the conclusion is false in order to be sure that the argument is invalid. It is not necessary to know this as a matter of fact, for its form alone reveals the status of the argument (see below, *Types of Deductive Argument*).

Accepting an argument as deductively valid does not of course mean that the conclusion is necessarily true. This will be the case only if the premises are also true. There are three possible combinations of truth values in a valid argument (one, for example, with the form common to those given above concerning Socrates and Cleopatra). Substituting the zoological terms used already, we have the first combination, that of true premises and a true conclusion, thus:

All snakes are reptiles.
All vipers are snakes.
Therefore all vipers are reptiles.

Alternatively we may have false premises (some or all) and a true conclusion as in:

All men are reptiles.
All snakes are men.
Therefore all snakes are reptiles.

Or, finally, there may be false premises and a false conclusion, as in:

All men are immortal.
All snakes are men.
Therefore all snakes are immortal.

The first of these combinations provides few problems, but the other two are worth a brief comment. It is commonly believed for instance that false premises must lead logically to conclusions that are false. However, a less extreme example than that given above (but still of the same form) readily shows this to be a fundamental mistake, thus:

All bats are birds.
All birds are flying vertebrates.
Therefore all bats are flying vertebrates.

The relation that holds between the premises and the conclusion necessitates that the latter cannot be false *if* the former are true (despite the fact that they are actually false). Equally of course it is important to realize that a true conclusion that is implied by a false premise is not proved to be true in a case such as this; merely that a conclusion is not automatically false because it is derived from falsehoods.

An error of the above type may result from bad analogy with the correct principle that a valid deductive argument yielding a false conclusion must rest on false premises. In daily life it is very common to hear arguments which draw logical conclusions from false premises. But this is also common in science where we are frequently confronted with competing hypotheses (premises) which cannot both (or all) be true. The value of a false hypothesis in scientific method is that it provides a basis from which we can validly deduce conclusions which are contrary to observed fact. We can thus eliminate the hypothesis because it is impossible for it to be true if the conclusion is false.

Types of Deductive Argument

The concept of form is of great value in determining the validity or invalidity of deductive arguments. One of the best ways of revealing an invalid argument is by what is known as the method of counter-example. Here the argument to be tested is compared with another argument having the same form, but in which the premises are known to be true and the conclusion false. For example, take the argument:

All Englishmen are British.
All Yorkshiremen are British.
Therefore all Yorkshiremen are Englishmen.

This has the form:

All As are Cs.
All Bs are Cs.
Therefore all Bs are As.

Note that this is the same form as that of the invalid argument given on p. 16 and which presented little difficulty. In the present case however the words substituted lend the argument a spurious conviction, which is of course precisely why the underlying form is of greater importance than any particular argument using that form. Any confusion over this new example is quickly dispelled if we substitute 'Scotsmen' for 'Englishmen'. Now we can readily see that we have an argument in which the premises are true and the conclusion false. This would therefore be a counter-example showing that the form, and hence the first argument, is invalid.

Another form of argument commonly encountered in science is that known as hypothetical or conditional. This is of the type 'if . . . then', for example, 'if the wind blows then the flags will fly.' In such propositions, the part introduced by 'if' is called the antecedent, and that introduced by 'then' is the consequent. It is important to realize that their use does not imply doubt. Although we may doubt that 'the wind blows', we have in fact no doubt that '*if* the wind blows then the flags will fly.'

A valid form of conditional argument known in logic as 'affirming the antecedent' runs as follows. When we assert the hypothetical proposition, 'if the rate of inflation increases, then prices in the shops will rise' and also assert its antecedent, 'the rate of inflation increases', we must necessarily infer the truth of its consequent. The form of this argument may be represented simply as:

If A, then B.
A.
Therefore B.

The importance of this procedure lies in the fact that it allows us to assert the consequent of a hypothetical proposition indirectly, when by direct means we can assert only the truth of the whole proposition and of its antecedent.

A common fallacy is that of 'affirming the consequent'. Thus, suppose we know that 'if the rate of inflation increases, then prices in the shops will rise' (if A, then B) and also assert that 'prices in the shops have risen' (B). We cannot in this case necessarily infer that 'the rate of inflation has increased' (therefore A). The hypothetical proposition asserts only that if the antecedent is true the consequent is true. It does not say that the consequent is true on the exclusive condition that the antecedent is true. Quite clearly, the rise of prices in the shops may be the result of something other than an increase in the rate of inflation.

This leads to an interesting third example which is a valid means of disproving a suggested hypothesis. The method of 'denying the consequent' goes thus:

If the rate of inflation increases then prices in the shops will rise.
Prices in the shops do not rise.
Therefore the rate of inflation has not increased.

The form of the argument is:

If A, then B.
Not B.
Therefore not A.

The hypothetical proposition says that it is not the case that the antecedent is true and the consequent not true.

The final argument of this series is the fallacy of 'denying the antecedent'. We assert that 'if the rate of inflation increases, then prices in the shops will rise' (if A, then B) and that 'the rate of inflation does not increase' (not A). May we then infer the falsity of the consequent (not B)? We may not, for the reason given in connection with the previous fallacy; we have not excluded the possibility that changes in inflation and prices may occur independently.

One other form of deductive argument is important. This is the well-known '*reductio ad absurdum*'. The principle here is that we assume as false a proposition that we actually wish to prove true (assume not A for A). By a valid argument we then deduce a conclusion that is false (one which contradicts not A), showing that the assumption (not A) must therefore have been false also. If the assumption is false, our original proposition (A) must have been true.

To take a trivial example, suppose that for some reason we feel uncertain about the truth of the conclusion in the following syllogism, yet know the premises to be true:

Some snakes are not vipers.
All snakes are reptiles.
Therefore some reptiles are not vipers.

If we begin by assuming that the conclusion is false, its contradictory form, 'all reptiles are vipers' must be true. On this basis we construct a new syllogism, thus:

All reptiles are vipers.
All snakes are reptiles.
Therefore all snakes are vipers.

But the conclusion of this syllogism contradicts the initial (true) premise of the first syllogism. Since we also know 'all snakes are reptiles' to be true, the initial premise of the second syllogism, 'all reptiles are vipers' cannot be true. Thus the conclusion of the first syllogism, 'some reptiles are not vipers' cannot be false; it must therefore be true.

This may seem an elaborate and artificial procedure, yet it is one that can on occasion be extremely effective.

INDUCTION

So far we have considered the most rigorous form of logic in which a valid argument based on true premises necessarily yields a true conclusion. With inductive arguments there is no such compulsion in the reasoning and no logical inconsistency in accepting the truth of the premises while denying that of the conclusion (the central problem of induction).

A simple example would be the argument:

All men thus far have proved to be mortal.
Therefore all men are mortal.

The argument is sound, the premises true, but the conclusion does not *necessarily* follow. In this type of argument there are clearly grades of strength and it is for this reason that we speak of correct or sound inductive arguments, not of (absolutely) valid or invalid one. In the case above, there is a very high degree of strength or of probability. However if we had been examining, let us say, right-handedness, and 'all the men thus far examined' meant only ten men, or even one hundred, we should have a great deal less confidence in the generalization 'therefore all men are right-handed'. The degree of probability in a case such as this may be increased by new evidence ('new' in the quantitative sense of 'more', or in the qualitative sense of 'better'), and since the purpose of all arguments is to yield true conclusions on the basis of true premises, in induction we typically strive to achieve the highest possible degree of probability. (Problems associated with induction are further discussed in chapter 4, especially in the section on *Inductivism and its Problems*).

Types of Inductive Argument

The question of the accumulation of new evidence leads us to the commonest type of inductive argument, that by enumeration. Here we draw a general (universal) conclusion from a limited number of specific (singular) instances. Whereas with deductive arguments we draw out a conclusion already present in the premises, here we are putting in a number of instances which have not in fact been examined. That is, we are generalizing, or expanding the content of the premises into the unknown.

The form of the two arguments given above is the same, although the probabilities of their respective generalizations are of course widely different. We might express the form itself as:

P per cent of the examined sample of As are Bs.
Therefore P per cent of As are Bs.

In practice the value of P varies: in the first case (concerning mortality) it would be 100 (i.e. 'All As are Bs'), but in the second case (concerning right-handedness) it might be of the order of 80 (80 per cent of As are Bs), and we should accordingly speak of the two generalizations as universal, and as probabilistic or statistical, respectively.

There are countless examples of induction by enumeration in science and in daily life. A typical one might concern a claim for the alleged benefits of some new drug in cases of arthritic pain. There are many ways in which such a proposition could be tested. Clearly we cannot hope to test the drug on all people suffering from arthritis, so the best we can do is to give the drug to a selected group of patients and see whether they report a significant improvement which cannot be accounted for by chance. Such a trial could be quite elaborate, by introducing sophisticated precautions such as the use of placebos (dummy medication, say, sugar or salt, made up to be indistinguishable from the drug proper) and double-blind techniques whereby neither doctors nor patients know what they are giving or receiving. But no matter how carefully the trial may be planned and carried out, we are still relying upon generalization. If we conclude that the drug has beneficial effects on arthritic pain we shall be doing so on the basis of a limited sample only. Of course, the more representative the group, the more trust we may place in the conclusion. Thus the strength of the conclusion would be increased if we were able to test the hypothesis in a completely separate group of patients, perhaps in another country where climate, diet, and other environmental factors might be quite different.

In science we typically represent the degree of probability of an inductive inference by applying statistical analysis. Ideally, this provides a quantitative dimension. Thus in our investigation of right-handedness, a large sample of observations would enable us to improve on a vague generality of the type, 'the next individual examined will probably be right-handed.' It would enable us to predict that the probability of his being right-handed is, let us say, four-to-one.

The commonest fallacy in this sort of procedure is that of generalizing from insufficient data. Unfortunately of course, what constitutes an insufficiently large sample is often difficult (or impossible) to determine, but with hindsight we should be inclined to judge that the fallacy had been committed if the conclusion were found to be wrong. A well-known and disastrous case of this general kind was the thalidomide tragedy (where foetal deformities resulted from use of a sedative drug during pregnancy) in which there were not only too few observations, but also too few of the right kind. The observations were inadequate both in quantity and in quality.

The problem of the quality of evidence raises a number of issues which are not strictly the concern of the logician, for they have more to do with the trustworthiness of the 'facts' than with the objective status of the inference. Yet they are related to logical problems and well worth mentioning. In cases such as the thalidomide episode – indeed in the majority of medical research – it is standard practice to derive conclusions by analogy; thus experiments are performed on one species and the results extrapolated to another. Almost any drug trial that had reached the clinical stage would have been preceded by a trial of some sort on laboratory animals. How far can such evidence be trusted? The answer to such a question will depend upon how good the analogy is. With thalidomide the analogy was poor, not so much because of any species differences as such, but because the studies on pregnant animal species were poorly followed through.

Judgements as to the status of an analogy, as well as many other judgements, often involve what is known as an argument from authority. For instance, if the results previously reported in a given field by Professor Brown have established him as its leading figure, his views may be quoted in order to support some new findings. If his reputation is great, his authority may be held to extend beyond the area of his particular expertise, much as has happened with the opinions of Linus Pauling in the debate about the alleged prophylactic benefits of vitamin C in relation to the common cold. In this case Pauling's special authority is apparently based on the fact that he has won the Nobel Prize on two occasions. Critics of his views on vitamin C have tried to turn the tables by using another argument. They imply that Pauling's writings on this subject are of little significance because his Prizes were gained in quite different fields of scientific research.

Finally we should consider arguments which concern the idea of cause and effect. Not only are they common in science, but they also commonly involve the principle of generalization. When we take an aspirin for a headache and the headache shortly disappears, we naturally tend to infer that the action of the aspirin was the cause of our improved condition. But we cannot of course prove that the headache might not have ceased for some reason unconnected with the aspirin. If pressed, we should argue that the analgesic properties of aspirin are the well-known result of countless experiments and that our experience merely adds weight to the probability that it was the cause of our relief. Strictly speaking though, this is the fallacy called *post hoc ergo propter hoc* ('after this, therefore because of this'): it is not necessarily true that because B followed A, there was a causal relation between them. While it may be highly probable that this was the case with the headache and the aspirin, we may feel a great deal less certain when we hear, for example, that deflationary measures in the economy led to a surplus in the balance of payments.

There are two causal fallacies which are common and important in inductive logic. The one known as the fallacy of the common cause is that which notes two phenomena and interprets one as the cause of the other, while ignoring a third phenomenon which is the cause of both. For instance, it is often postulated that the portrayal of violence on television is responsible for the aggressive attitudes and behaviour of some people. However, we cannot easily rule out the possibility that there may be some deep-rooted condition in our society, or in human nature, which tends to promote both activities. If this interpretation were correct, then television violence would be merely a symptom of the underlying cause, not the cause of the aggressive behaviour. Finding the real cause in a situation such as this is obviously of critical importance both as a matter of logic and as a matter of fact.

The second fallacy is that of confusing cause and effect. Suppose we learn that the productivity of one factory is twice that of its competitor. The former's employees are industrious and they also drive to work in smart private cars. The latter's are lazy and also ride bicycles. If we identify the possession of cars as in some way responsible for the energy and dedication of their owners while at work, we may hope to eliminate the difference between the factories by presenting each of the second factory's employees with a new car.

The principle of induction may therefore be summarized by saying that the probability that a generalization is true is related to the quantity and quality of the evidence available to support it. But it is vital to understand that no proposition derived in this way can ever be conclusively verified. Such, in the last analysis, is the status of every scientific law or theory. Because the sun has risen on every morning in human experience does not *logically* prove that it will rise tomorrow. As the philosopher David Hume (1711–76) stressed, it is impossible to verify that the laws of causality will operate in the future as they have operated in the past, and this was why he concluded that science cannot be justified in the strictly rational sense at all (see also the section on *Inductivism and its Problems* in chapter 4).

LOGIC AND MEANING

We have now examined in an elementary way a few of the commoner argument forms which natural and social scientists encounter and to which the standards of logic may be applied. In all this we have assumed that the words used have actually been understood. The meanings of some of them, such as 'Socrates', 'man', and 'mortal' were of no special significance so far as the logic was concerned, because deductive arguments can be expressed symbolically. However, we said in considering the various types of statements used in constructing syllogisms that the other words, such as 'some', 'all', and 'no' had critical significance. Even in inductive arguments, which lack this precision, it is vital that exact meanings be attached to words and that these be applied consistently. The ordinary everyday use of language does not require a high degree of accuracy, but in logic and in scientific prose ambiguity and imprecision must be kept to a minimum.

The origin of the meanings of words lies in the learning situation. A child learns by repetition of experience to associate the sound of a spoken word (the symbol) with the object or the action to which it refers. Although as adults we can also get to the meaning of a word by defining it with other words, eventually some of the same words are bound to be used again so that the process becomes circular. In any case, words and sentences can be used for many purposes other than for strict reference to objects or actions, as for example in poetic allusion or political rhetoric. But so far as logic is concerned there is no place for the persuasive or deceptive use of language. Logic takes the ordinary meanings of words for granted – indeed, in deductive logic, it does not need to assume any given meaning for a word, except in the case of those such as 'some', 'all', 'no', etc. – and examines the relations between propositions on this basis alone. If we remember this, especially when trying to analyse ordinary discourse which cannot readily be broken down into simple logical form, it is possible to circumvent many of the more trivial, yet time-consuming, interactions between logic and language.

For what it is worth, we may illustrate the kind of statement that has its logical fascination but is of no value in the context of science, by the famous, if somewhat indulgent, 'Paradox of the Liar'. If someone says, 'I am lying', we cannot readily say what would make the claim false. But since he says that he is lying, what would make his claim true is if he were actually telling the truth. In that case, what he says must be true, and since what he says is that he is lying, if he is telling the truth then he is in fact lying. Thus we also cannot say what it would be like for his claim to be true, because that leads to a contradiction.

If what he says cannot be true, then he must be lying. But to say that he is lying is to say that what he says is untrue. So when he says that he is lying, this must be untrue. But if it is untrue that he is lying then he must be telling the truth. Thus if we assume that he is lying we conclude that he is telling the truth, which is a second contradiction!

All that can be understood in such a case is that whichever way we take it, the statement tells us nothing. Fortunately, it seldom serves the interests of scientists to indulge in such word-play and it is generally possible in scientific prose to identify the essentials of an argument and to separate its component parts as a prerequisite to logical analysis. Nevertheless it is common to find that conclusions and premises occur in sequences which differ radically from those of logical conventions; that meaning is obscured by inaccurate or excessive use of words; or that an argument is incomplete owing to the absence of some premise that is only implied. When, for instance, we assert that one molecule of oxygen combines with two molecules of hydrogen *because* the atoms of oxygen molecules are divalent and those of hydrogen are monovalent, it is important to realize not only that our assertion is actually the conclusion of the argument, but also that the first, or major, premise, 'a simple divalent atom will combine with two monovalent atoms,' has been omitted.

The social sciences in particular have much to gain from a logical approach to the analysis of argument because the data with which they have to deal are typically more complex than those of the natural sciences (see chapter 5). A good example might be the assertion which was briefly mentioned above, namely that the aggressive behaviour of some young men is attributable to the violence that they watch on television. As we have already seen, this is likely to be a difficult case to prove for we might, among other hazards, be in danger of falling for the fallacy of the common cause.

Without going into detail, let us see if the assertion would be any easier to disprove. For instance, would a list of young men who exhibited aggressive behaviour but did not watch television, serve as a counter-example? In order to make decisions of this sort, it would first be important to identify the matters of fact and the matters of logic involved. Any inductive generalization that may be drawn from instances where aggressive young men do or do not watch television would be concerned with 'factual' relations; in principle the accumulation of more evidence may be expected to enhance the strength of one case or the other. However, the question as to what sort of evidence would confirm or falsify the original assertion concerns more than mere 'facts' (difficult as they may be to pin down). It also concerns the *meaning* of the assertion itself. Thus a list of aggressive young men, say from history, who were not influenced by television, would be a counter-example to the assertion that if a young man behaves aggressively this must be a result of his having watched violence on television. Such a list, we could say, would contradict the assertion that the watching of television violence is a *necessary condition* for the production of aggressive behaviour. However the list clearly would not be a valid counter-example to the assertion that young men who watch television violence act aggressively. The assertion here says that the television violence is a *sufficient condition* for the production of aggressive behaviour, but this does not exclude the possibility of other causal factors. A counter-example to this particular interpretation would therefore be a list of young men who were addicts of television violence yet who did not act aggressively.

In reality of course it would be easy to find counter-examples to refute the original assertion in both its meanings, and it would be important to see any causal connection between television violence and aggressive behaviour as having, at most, some level of statistical probability rather than as being direct and categorical. Nevertheless, this example shows in an elementary way how ordinary discourse may be better understood by careful analysis. More complicated and protracted arguments can be treated in much the same way, not only revealing their weaknesses, but also showing how they may be strengthened. In general we can say that logic provides us with tools which are necessary to untangle the conflicts and incompatibilities that we all experience in our own beliefs. By drawing out the deductive inferences from propositions that we think we subscribe to, logic may lead us to accept viewpoints which upon superficial acquaintance might seem surprising. And equally, by providing methods for the weighing of evidence, logic may sometimes encourage us to reject propositions which were hitherto held to be self-evident premises.

Finally in this section it is perhaps necessary to say a brief word about the relation between logic and psychology. The confusion that is still occasionally to be found between the two stems from the old and ambiguous definition of logic as the science of the laws of thought. Whereas the psychologist is interested in the nature of the mental processes that occur in thinking, in the sense of the 'processes going on inside a person's head', the logician is concerned with them only in so far as they provide methods of testing arguments or inferences. In this connection logic is no different from any other branch of intellectual activity, where problems are analysed without contemplation of the actual mechanisms of the brain which are associated with the analysis. The sorts of problems analysed by modern logic and the processes used in the analysis are, in fact, closely similar to those of mathematics and it is in this sense that it is sometimes said that thought is no longer necessary in logic. This view is of course not intended to imply that logicians and mathematicians do not think, but is derived by analogy with modern electronic machines which analyse problems by using 'computer logic', a binary system with only two values, say 'on' or 'off'. In deduction the two possible values are validity or invalidity. Provided that an argument, however complex it may seem, can be broken down into a series of sub-arguments, each of which can have only one of these two possible values, the problem can in principle be resolved by a suitably constructed and programmed machine. (Needless to say, the machine can give no guidance whatever as to the truth or falsity of the information – that is, the premises – fed into it.)

While some appreciation of the possibilities and limitations of logical processes may help to reinforce the contention that logic has no special relations with psychology, it is important to remember that there are many contexts in which so formal and rigorous a system as deduction cannot be applied. In such cases, the application of logical principles of the more general type outlined above remains the only way of satisfactory analysis.

3

The Scientific Attitude: Science in Practice

The word 'science' is used in two different ways. On the one hand it refers to a body of knowledge, and on the other to a set of rules by which this knowledge is to be collected. In neither sense is the meaning of the word perfectly clear-cut and it would certainly be a mistake to pretend otherwise. But it would be equally misleading to pretend that there is not a fairly widespread consensus of opinion as to what science really is. It is the purpose of this chapter to identify the principles which, in addition to those of logic, seem almost universally to be associated with the conduct of scientific enquiry. These principles together constitute what is often referred to as the scientific attitude and we shall try to illustrate the special flavour of this by means of cases from history. To begin with, however, it is useful to set out the essential features which distinguish the body of knowledge we call science, for we shall then be in a better position to summarize the simple practical formula for building it up, which runs through all the historical studies.

GENERAL CHARACTERISTICS OF SCIENCE

A dictionary definition of science might be something along the following lines: 'Knowledge of the real world ascertained by observation, critically examined and classified systematically under general principles.' In broad, if somewhat idealistic terms, we can say that this knowledge should provide an explanation of what is of value in past discoveries and it should also make some prediction of future events. Furthermore, it should promote scientific research (new discovery) by providing concepts which give a sense of understanding of the causes of events in the world and which help to communicate this understanding to others. Scientific knowledge should be universal in the sense of independent of space and time; it should be presented explicitly so as to be intelligible to all qualified practitioners; and it should have empirical relevance such that all can evaluate the correspondence between its theories and their practical implications.

If the many different ways of accumulating scientific knowledge are nevertheless recognizably of a kind, they must all share certain common characteristics. There is a good deal of argument about what these characteristics are, let alone about what they should be, but undoubtedly the most universal in addition to the soundness of logic, is 'objectivity'. We all know roughly what is meant by 'being objective' and for the moment that is quite sufficient. In due course we shall have to look in some detail at the philosophical limitations of objectivity (notably in the section on inductivism in chapter 4), but in passing it is worth noting that the high status accorded to the attempt to be objective seems to be confirmed by the growing role it now finds outside the laboratories and formal meeting places of the scientific community. Objectivity is increasingly valued in those many areas of modern life where evidence needs to be assessed as a basis for some decision-making process; we need think, for instance, no further than the ever-expanding field of 'management studies', in which the aim is to make commercial enterprises more 'efficient'. This growth of the scientfic attitude has important implications, and the far-reaching problems for society which some observers see as resulting from its intrusion into human affairs are discussed in chapter 7.

Logic, together with pure mathematics, differs from the disciplines normally included under the term science by being 'non-empirical'. This expression simply means that logical and mathematical propositions are demonstrated independently of empirical evidence; they make no reference to things or events outside of the human mind, depending upon postulates but not upon observations. All of the other disciplines regarded as scientific, covering the wide spectrum from sociology and economics, through anthropology, psychology and biology to the physical sciences such as physics and chemistry, are 'empirical'. Their propositions depend ultimately on the evidence of the senses.

One of the characteristically objective features of observation in the empirical sciences is that it places special emphasis on those aspects of things and events in the world which can be agreed by all observers; that is, it tries to achieve maximum consensus (see chapter 1). This accounts for the particular interest in quantitative properties, those properties which can be measured. Precise measurements, especially those performed by impersonal instruments, are matters of widespread agreement, whereas the qualitative (perhaps aesthetic or subjective) properties of things involve personal judgements and commonly lead to dispute.

The types of observation which are made will naturally depend upon the discipline in question. Scientific techniques range from the surveys of social research and the collection of ancient relics in anthropology, to the elaborately contrived experimental measurements of high-energy physics. But despite the enormous differences between the observational techniques of the various sciences, it is possible to argue that the fundamental methodology is the same in all. While for reasons of convenience we often refer both here and elsewhere in the book to the conventional distinction between the 'natural' (i.e. the physical and life) sciences and the 'social' sciences, we are concerned in what follows with disciplines across the whole spectrum.

It was suggested earlier that what the empirical scientist actually does in practice is to put forward ideas which he can test (roughly speaking for truth or falsity) by the observations of his senses. This reciprocal interaction between the mental process of thinking up ideas and the physical process of testing them is the kernel of the scientific method. New ideas occur to the scientist within the two-fold context of his training and of his experience. The former has given him the principles and the factual framework of his subject, and the latter provides certain expectations as to what is likely to be right or wrong. Consciously, or often unconsciously, the scientist examines premises and tests arguments. New hypotheses are compared in terms of their implications, first against the array of relevant factual observations already available, and then against a new body of observations collected specifically for the purpose. Ideally, this process of collection will constitute a particular type of controlled observation called an 'experiment', in which an attempt is made to eliminate as far as possible all factors likely to obscure the crucial data to be collected. Thus simplifying conditions are established in order that the system to be observed may be controlled, manipulated and thereby understood. When experiments of this nature are not possible, as is still the case with many of the biological and social sciences, the scientists will make extensive observations on his world 'as it is', measuring relevant variables and ordering his data according to some appropriate system of classification. Comparisons may then be made in contrasting circumstances and, where clear and simple relations are not forthcoming, grades of probability may be established by statistical analysis.

There are numerous historical episodes during which, intoxicated with the excitement of discovery, scientists have seemed to abandon all standards of objectivity, but it is important to realize that any such emotional behaviour, while it may reassure us that scientists are, after all, only human, is not that part of the overall enterprise which marks it out as scientific. Emotional, even irrational, behaviour is rather to be seen as that part which science shares with all creative activities. As such it is concerned, just as other such activities are concerned, with the *discovery* of new ideas.

What needs emphasis is that the characteristically scientific part of the whole enterprise is, by contrast, concerned not with discovery as such, but with the *justification* or testing of the ideas that have been discovered. In science new discoveries stand little chance of acceptance unless presented in a manner which excludes dogma or mere opinion that is unsupported by independently testable evidence. Consequently there is less scope for the involvement of the 'self', meaning the personality and prejudice of the individual practitioner, than there is, for example, in art or literature. Even the ideas of a creative 'genius' among scientists must still be subject to the public scrutiny of his fellows before being accepted into the body of scientific knowledge.

There is another and related respect in which science also differs from other forms of scholarship, such as the study of literature or philosophy. While in philosophy, for example, it is still possible to spend much of a life-time analysing the writings of the Greeks (Whitehead was only partly in jest when he described modern philosophy as nothing but 'footnotes to Plato'), it is a contradiction to speak of a modern 'Platonic' or 'Aristotelian' scientist. The reason for this has to with the feeling that science somehow 'progresses' in a way which sets it apart from other disciplines. For instance, if Plato or Shakespeare returned to review philosophy and literature since their time, it is by no means certain that they would acknowledge that 'progress' had occurred. But for Galileo, Lavoisier or Darwin, there would be no doubt that physics, chemistry and biology had advanced in a very obvious sense, and they would be likely to see their own work as steps in the direction of that advance. No one would argue that Plato's philosophy or Shakespeare's drama have been 'surpassed'; indeed, they are often referred to as 'timeless', which is intended to mean that their value and relevance today are as great as they ever were. But in science a good many theories, even of the nineteenth century, are now regarded as partially or wholly inadequate for explanatory purposes, because they no longer 'fit the facts'. We might say therefore that science 'moves on' while other disciplines simply 'move about'. (It is hardly necessary to add that the gradual 'dating' of scientific theories in no way diminishes their importance for the historian, nor indeed for the modern scientist who seeks a perspective broader than that exclusive to his own generation.)

SCIENTISTS AT WORK

Having now introduced the basic conception of the scientific attitude as the objective testing of hypotheses by observation, it is time to illustrate this formula by some actual examples from the history of science. Cases from earlier times are generally more useful in this respect than those taken from the twentieth century because the ideas are technically less difficult, even when they have not already been absorbed as a part of common knowledge. With the advantage of hindsight, it is also easier to see the essential structure of the conceptual and experimental processes involved, and the following accounts are therefore presented quite consciously through twentieth-century eyes. As a result they do not claim to be 'good' history in the modern sense which seeks empathetic understanding, and (particularly in the case of Harvey) they give little indication of the subtle forces which may influence a scientist's overall philosophical position. This, however, is not their purpose. What the case histories do provide, is a quick and informal insight into the essential nature of scientific method, by concentrating on the attitude adopted by famous individuals confronted by particular problems.

William Harvey and the Circulation of the Blood

The first example of the practical application of the scientific attitude concerns work which has been regarded by many modern authors as a methodological classic. It is empirical and quantitative in character; it illustrates the interaction and interdependence of theory and observation; and it appears to exemplify the precedence of hypothesis over test. Moreover, it shows how the scientist must ruthlessly reject an old theory, no matter how well established, as soon as it no longer finds support in new evidence.

William Harvey (1578–1657) studied medicine in the Italy of Galileo's time. He later returned to England, became a Fellow of the Royal College of Physicians, and then personal physician to James I and Charles I. King Charles gave him access to the Windsor deer parks for his experimental work and Harvey examined over forty species of animals, many of them cold-blooded, such as fishes and reptiles, as he developed his interests in the activity of the heart.

Harvey had inherited the enormously influential doctrines of the Greek physician Galen on the workings of the body, which taught that the venous blood was manufactured by the action of the liver on the contents of the intestine. This blood was distributed throughout the venous system by means of an ebbing and flowing motion. Some of it entered the right side of the heart where its impurities were taken off into the pulmonary artery for exhalation from the lungs. A small volume of venous blood from the right ventricle was alleged to pass through minute pores in the septum separating the two sides of the heart. Once in the left ventricle the blood met with the spirit or pneuma which had itself entered the heart from the outside air by way of the trachea, lungs and pulmonary vein. The dark venous blood then became transformed into light arterial blood which was distributed throughout the arterial system (figure 1).

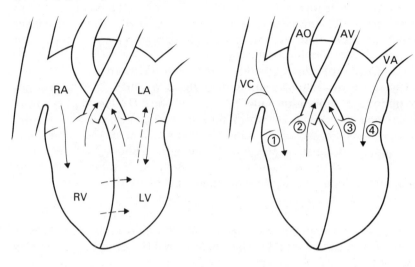

AO	aorta	1	tricuspid valve
AV	'arterial vein' or pulmonary artery	2	pulmonary valve
LA	left atrium	3	aortic valve
LV	left ventricle	4	mitral valve
RA	right atrium		
RV	right ventricle		
VA	'venous artery' or pulmonary vein		
VC	vena cava		

Figure 1   *Diagrams of the blood flow through the heart according to Galen (left) and Harvey (right)*

Harvey set out to examine the 'motions and uses of the heart', and sought to discover these from actual inspection, and not from the writings of others. He soon observed that the contraction of the left ventricle was followed immediately by an expansion of the great arteries, and reasoned that this must be because blood passed out of the former into the latter. His dissections of the heart revealed valves, the arrangement of which was such as to allow the passage of blood in one direction only. For example, they prevented the backflow of blood into the right auricle upon contraction of the right ventricle. The implication again was that blood must pass from the right ventricle into the pulmonary artery.

Harvey also demonstrated the existence of valves in the veins which permit blood to flow only in the direction of the heart. The simple 'experiment' of severing a vein was always followed by a trickle of blood towards the heart, but severing an artery resulted in vigorous pulsations away from the heart and the speedy death of the animal. Harvey here reasoned that death resulted from the failure of the blood ejected from the arteries to return to the heart by the veins. The experiment confirmed his hypothesis that 'there might be a movement, as it were, in a circle.'

Observations of this sort were more than sufficient to give Harvey the insight to formulate his famous hypothesis by 1615. Yet he did not commit it to publication until 1628 in his book *On the Motion of the Heart and the Blood,* by which time he had confirmed it by further observations. Perhaps the crucial test, that which made the whole study so simple and classic a precursor of modern method, was the one in which he performed quantitative calculations. Estimating that the amount of blood 'extruded by the individual beats' of the human heart to be at least half an ounce, he calculated that 'in half an hour the heart makes over a thousand beats . . . [and] five hundred ounces . . . [have] been passed through the heart to the arteries, that is . . . blood in greater amount than can be found in the whole body.' His hypothesis therefore explained his own observations and gained support from his calculations which, at the same time, led to the rejection of Galen's authoritative doctrine. The latter was contradicted by the demonstration of valves in the veins, and could hardly be claimed to account for the enormous output of the heart.

We may see Harvey's demonstration of the non-existence of pores in the thick intraventricular septum (that which he could not see he naively believed could not be there) as a simple example of the valid logical form, 'denying the consequent'. Thus, it may be reasoned, if Galen's hypothesis that the blood in the left heart has entered directly from the right heart is true, then there must indeed be demonstrable pores in the septum (if A, then B). But the evidence, in the form of a thorough examination of the septum, reveals no such pores (not B). Therefore the hypothesis is not true (therefore not A).

Harvey never did identify the one missing link in his circle, namely the existence of the blood capillaries linking the smallest arteries to the smallest veins. (The first observation of the capillaries was made by Marcello Malpighi, shortly after Harvey's death.) This crucial omission raises the important point that whereas his inability to identify the capillaries was not taken by Harvey as reason sufficient to reject his hypothesis of circulation, his failure to find pores in the septum was seen as one critical element leading to the rejection of Galen. It might, after all, have been that the septum was examined with instruments inadequate for the demonstration of pores that did in fact exist. That is, the second argument is true only if its two premises are true. But of course in the strict deductive sense the confirmation of Harvey's hypothesis does not follow simply as a result of its implication (derived from his observations) being found to be true. Any such conclusion would be committing the fallacy of affirming the consequent (if A, then B; B; therefore A). So when we say that Harvey's empirical evidence confirmed his hypothesis, this can be true only in the limited sense that evidence in inductive arguments may be graded in importance. Nevertheless, no one doubts its reliability in practice, and the whole of modern physiology is founded on the resulting conception of the blood as the body's transport system.

Stephen Hales and the Movement of Sap

Stephen Hales (1677–1761) who spent much of his life as curate of Teddington – a fine example of the English clergyman-natural philosopher of the eighteenth century – studied in Cambridge while it was profoundly under the spell of the Newtonian methodology. This is clear in his own attempt to find mechanistic explanations for the processes of living things. This case history therefore illustrates in particular the significance of such influence, together with the importance of new apparatus which can initiate a technical breakthrough and provide accurate measurements.

Hales's classic book on plant physiology, *Vegetable Staticks* (1727) must also be attributed in part to the impact of Harvey's theory on the circulation of the blood, for it was in a fashion analagous to Harvey's that he set out to investigate the dynamics of the movement of the sap. That there *was* movement had of course been widely understood ever since the time of Aristotle, but Hales was the first to tackle the problem by quantitative experiment.

First he wanted to know the amount of fluid involved and the velocity at which it moved. He determined the quantity transpired by carefully weighing a cabbage plant over several days and calculated the rate of rise of the sap in the stem by means of measurements of its area. The area of the leaves was much greater than that of the roots and from this he estimated that the movement of the sap must be nearly eleven times faster in the latter. He was deeply impressed by the dependence of plants upon water and calculated, for example, that bulk-for-bulk, a sunflower needs some seventeen times as much as a man.

At some time Hales must have formulated the hypothesis that it is the leaves which are essential for the movement of water through the plant, for he performed experiments apparently designed to test it. These showed that the quantity of water imbibed by a branch was 'more or less in proportion to the quantity of leaves they had' but they said nothing as to the mechanism by which they exerted their effect. Accordingly, he took a long glass tube into which he inserted a branch so that he was then able to measure directly the rate at which water was imbibed. When the leaves of the branch were immersed in water, thus preventing transpiration, little fluid was imbibed, but on exposing the leaves to the air even when the branch was held upside down, fluid movement in the tube was rapid. The fluid movement in the stem was thus *caused* by transpiration from the leaves. But while it was clear that water was available to the roots of the plant in the soil, Hales had still to show that the fluid transpired through the leaves is also water. For this purpose he enclosed the leafy branches of a variety of plants in chemical retorts and collected the clear liquid given off. By tasting this, and by measuring its specific gravity, he satisfied himself that it was indeed water, although not pure for 'when reserved in open viols, it stinks sooner than common water'.

Having further satisfied himself that there was sufficient water in the soil – as provided by rain and dew – to supply the plant, Hales now wanted to know where it went before transpiration through the leaves. His means for answering this question beautifully illustrates the importance of new techniques. He performed, in fact, the first of a kind of experiment that is now common in many branches of biology, that using a 'trace-element'. By providing water strongly scented with camphor, sassafras and various perfumes, he was able to show, by the testimony of his nose, that far more water had reached some parts of the plant than others. In the case of a vine, for example, the scented water 'did not penetrate into the grapes, but very sensibly into the wood and stalks of the leaves'. With apple and pear trees he also showed that the fruits were not affected.

By another ingenious technique Hales was able to measure the pressure developed by the sap in the plant. He designed his 'aqueo-mercurial gauge' (figure 2) from two glass tubes cemented together, the larger filled with water and attached to the stem of his plant and the smaller, also full of water, inserted in a bowl of mercury. The gauge then doubled as a reservoir from which water could be drawn and as a device for measuring the strength of the suction. (Hales even identified the limitations of his gauge, specifically noted a systematic error, and described the method of correcting for it.) With it, Hales confirmed his earlier findings on the importance of the leaves, for the height to which the mercury was drawn up the small tube was very much lower after these were removed, and he also showed that the power of imbibing water is less in needle-leafed (evergreen) plants than in broad-leafed (deciduous) ones. Stems also proved able to draw water from the gauge in either direction along their lengths, provided there were plenty of

Figure 2 *Hale's aqueo-mercurial gauge, here used 'to find out the force with which Trees imbibe moisture'*

leaves, and the pressure developed was unaffected by removal of the bark.

These last observations were probably surprising to Hales for it was widely considered in his time (perhaps by analogy with Harvey's work) that the sap moved up the plant in the centre of the stem and down again in the outer part. That is, it was thought to circulate. But Hales was too good a scientist to respect authority for its own sake, and when he immersed the leaves of a branch in water he was able to show that movement of fluid *towards* the leaves was slow, although substantial quantities of water were actually absorbed *through* the leaves. Furthermore, a branch from which the bark and the last year's ring of wood had been removed for three inches, was still able to imbibe water through the remainder of the stem. This showed that water *could* pass up through the centre. But if sap were returning between bark and new wood, the upper part of the wound should have been moist, whereas instead, it was white and dry. There cannot therefore be a circulation of sap. After further experiments Hales was finally convinced of the error of the old theory and concluded that 'the sap ascends between the bark and wood, as well as by other parts'. This view was consistent with his previous results and brought him to what is essentially the current position, namely that the upward movement of the sap through the whole area of the stem is the result of 'the strong attraction of the capillary sap vessels . . . whereby the sap is carried to the top of the tallest tree, and is there perspired off through the leaves.'

William Wells and the Formation of Dew

William Charles Wells (1757–1817) was an American Tory who came to England after the revolution. He published his famous *Essay on Dew* in 1814. In common with the work of Harvey and Hales, that of Wells is a valuable object-lesson because, with minor modifications from the original, it can be presented in the form of an 'ideal' treatise on the scientific attitude and yet it deals with subject matter familiar to everyone, and with concepts that all can understand.

Wells first set out clearly to define the problem. Instances where moisture appeared on objects exposed to the air, but in the absence of obvious water, were distinguished from instances where the deposits were the result of natural precipitation. Earlier ideas on the conditions necessary for the appearance of dew were then discussed. Aristotle, for example, had said that dew was a form of invisible rain that fell only on calm and clear nights, while Musschenbroek (1692–1761), a Dutch natural philosopher, claimed that metals, unlike other materials surrounding them, were often free of dew. From these previous accounts and from his own preliminary observations, Wells drew up a list of situations which typically lead to the formation of deposits of water, similar to that of dew. Among these he identified the familiar appearance of moisture on the inside of windows in cold weather, on interior walls when warm weather follows a long period of frost, and on mirrors that are breathed on. All of his examples had one feature in common, namely that the surfaces upon which moisture formed were always colder than the air in contact with them, and Wells had sufficient insight to abstract this simple connection and to express it in terms of his initial hypothesis.

Unlike his predecessors, however, Wells was not satisfied merely with the approach of the natural historian, but wished to test his hypothesis exhaustively by experiment. First he verified the hypothesis that dew formation occurs when cold objects come into contact with warmer air by measuring directly the temperature of the night air and of various materials observed. In every case the hypothesis received support and Wells accordingly induced (by enumeration) from his limited number of instances the general conclusion that this difference in temperature held in every case.

Here, however, it was possible that he had committed the fallacy of confusing cause and effect. While Wells considered that the coldness of the objects dewed was the cause of the deposition of moisture, it could equally be argued that this coldness was rather the effect of the deposition of moisture from the air. This possibility could be ruled out by further experiment, and Wells resorted to the careful weighing of objects before and after dew formation, and to further temperature measurements. Thus he soon found that some objects could be cooler than the air and yet free of dew; in these instances the deposition of moisture was not the cause of cooling. Although this observation did not of course prove that deposits of dew do not cause cooling (since some other cause might operate when dew was absent), it nevertheless did lend weight to the argument that the coolness of the surface is a cause, not an effect, in the case of dew formation. The absence of dew in some instances might then be accounted for in terms of the dryness of the air (an explanation which happens to be correct).

One of Wells' most remarkable series of investigations aptly illustrates the desirability of setting up simplifying conditions in experiments; the scientist may then examine the influence of one variable at a time so as to isolate its influence from those of other variables in the system as a whole. Leaving all other circumstances the same, he looked at a wide variety of different materials in contact with the earth and found that the coldest on any given night were the poor conductors of heat (feathers, wool), while the least cooled were the efficient conductors (metals). With this discovery he was able to refine his hypothesis further and to present it in a much more sophisticated and quantitative form. It could be said not only that dew appears when cold surfaces contact warmer air, but that there exists a positive relation between the degree of coldness and the amount of moisture deposited. This hypothesis was tested by deducing consequences from it. For example, if a good conductor (which when in contact with the earth draws heat from it rapidly, remains relatively warm, and is therefore little dewed) is now suspended just above the earth by fine threads, it should be colder than when resting on the ground and consequently more heavily dewed.

This prediction proved correct and thus gave substance to Musschen-broek's observations by confirming that the thermal conductivity of a material was the underlying cause of differences in temperature and hence of the copiousness of moisture deposited.

Elaborate experiments with weighed samples of wool then showed Wells that the degree of exposure of objects to the night sky (another single variable) also affected dew formation. Wool suspended in the air above a wooden plate unfailingly gained more weight as a result of dew deposition than did wool suspended some way beneath the plate. This general effect proved constant in a host of different situations and confirmed the ancient observation that overhanging trees prevent or minimize dew formation. It also suggested that cover of any sort acts in the same way as clouds by reflecting back the warmth radiated from the earth at night, and prevents objects on and near the ground from becoming so cold and hence from collecting so much moisture. Aristotle was thus shown to be in part right and in part wrong.

Wells's logical position was typical of the problems posed in the testing of hypotheses in science, and it introduces us to the idea of auxiliary hypotheses that is discussed further in chapter 4. For example, take the hypothesis that the deposition of dew is caused by contact between a cold object and warmer air, and consider the test implication that a feather placed in contact with the ground will become dewed but a piece of metal will not. This implication clearly cannot follow deductively from this hypothesis alone; its derivation presupposes other hypotheses concerning the condensation of water from air and the thermal conductivity of different materials. Similarly, a test implication concerning the effect of cloud or other shelter on dew formation will presuppose a further premise about the radiation of thermal energy through space. Thus in the strict sense Wells could not assert that if his hypothesis were true the test implications must also be true, but only that if his hypothesis and the additional, or auxiliary, hypotheses were true then the implications must be true. As it happens they were (in the sense that we now appreciate), and Well's general explanation of the causes of dew formation remain largely unchallenged down to the present day.

Émile Durkheim and the Social Causes of Suicide

As a final case history we may examine the work of the French sociologist Émile Durkheim (1858–1917), whose book *Suicide* (1897), is perhaps as good an example as any of how scientific method can be applied to a complex problem of society which, however, offers not the slightest possibility of experimental manipulation. It is probably the first work in sociology in which the collection of factual information is clearly guided by hypothesis, and it is the perfect example of the application of Durkheim's equally famous classic, *The Rules of Sociological Method*. Durkheim thus tried to practise what he preached.

The initial hypothesis was controversial to say the least. It was that suicide, perhaps the most intimately personal action that an individual can take, is a phenomenon which is nevertheless to be understood, not in terms of individual psychology, but rather in terms of social forces wholly external to the individual. In fact suicide is a specific example of Durkheim's more general thesis that the subject matter of sociology ('social facts') cannot be reduced to that of psychology.

Recognizing that sociology as a whole was still at the stage of fact-gathering rather than problem-solving, Durkheim said that if he were to solve problems, the sociologist would have to investigate 'groups of facts clearly circumscribed, capable of ready definition, with definite limits . . . .' Accordingly, for his purposes the term suicide was used for 'all cases of death resulting directly or indirectly from a positive or negative act of the victim himself, which he knows will produce this result'. On the basis of this definition it might seem almost self-evident that the study of suicide *is* a matter for the psychologist, yet Durkheim insisted that the suicide in a society, when treated as a whole (i.e. as suicide *rates*), constitutes something altogether different from the mere sum of the individual suicides. Indeed, his revolutionary hypothesis had obviously come from a preliminary examination of the suicide rates for different countries, for in many cases he showed these to be more constant than the corresponding overall death rates.

First Durkheim set out to exclude a variety of possible extra-social causes of suicide. Take insanity, for example. If all those who commit suicide are *defined* as insane, then of course there is little case for a social enquiry; but this is unreasonable because there is an enormous difference between certifiable madness and the depression of an otherwise normal, balanced person. Yet both may commit suicide. Furthermore, the statistics provided no support for this psychopathic theory, for in several countries Jews had a higher rate of insanity than Christians but a lower rate of suicide. In a similar fashion, Durkheim considered other possible causes, such as alcoholism, geographical and climatic factors and 'imitation', subjecting them all to careful analysis and eliminating each in turn.

He then moved on to examine what he considered to be the genuine, or social, causes of suicide. At this point he reaffirmed that he regarded the conventional explanations for suicide, those in terms of family trouble, jealousy and so on, as likely to be no more than symptoms of underlying social causes such as religious, political and occupational affiliations. With religion, he tried to minimize possible cultural influences by examining statistics which both compared different countries and different communities within countries, finding that Protestants in all situations were significantly more prone to suicide than Catholics. The reason for this, he said, was not a difference of doctrine, since suicide is forbidden to Protestants and Catholics alike, but rather a difference in the cohesive force of the respective religious communities. The denominational nature of Protestantism meant that the individual was left much more to seek his own salvation in his own way, whereas the Catholic was expected to accept the one authority and to conform to a tightly integrated social organization which, however, protected him as an individual. It was thus in the religious *society* that the deterrence of suicide was to be sought.

It was likewise with the family. The suicide rate of the married person was universally lower than that for the unmarried, but the statistical evidence showed that the explanation for this lay, not in the advantages of marriage itself, but in the cohesion provided by the family society, that is, by spouse *and* children. It was this that explained why there was no relation between *marriage* rate and suicide rate; why the latter was higher among married men without children than among widowers with; and why it was lower among unmarried women than among childless wives.

From his studies of the effect of religion and the family (and of political groupings) Durkheim derived his more general theory that 'suicide varies inversely with the degree of integration of the social groups of which the individuals form a part'. Although we have considered his analysis only superficially, enough has been said to illustrate the fundamental contention that the problems of society (or at least some of them) are amenable to a method of enquiry that does not differ in principle from those methods employed in the natural sciences (see chapter 5). By restricting the scope of his investigation and by making explicit the problems inherent in it (especially those of definition and interpretation of evidence), Durkheim was able to formulate a theory which could be tested against objective evidence, which was so tested, and which successfully withstood the tests.

There is perhaps one further lesson to be gained from a study of Durkheim's book. It stems from the fact that Durkheim was writing at a time when sociology was struggling to establish itself as in independent science, and it is essential therefore to see it in the perspective of history. The problem was that Durkheim, having presented his theory together with the extensive evidence to support it, then turned to a fervent defence of the idea of the objectivity of social facts and resorted to an extravagent and somewhat rhetorical use of language which proved altogether counter-productive for sociology itself. By raising questions such as the 'reality' of social forces which, for him were 'things *sui generis* and not mere verbal entities', and by speaking of 'the collective force impelling men to kill themselves', he fed the prejudices of those who wished to denigrate his discipline as 'metaphysical' and unscientific. For the immediate purposes of his thesis on suicide there was no need to raise such issues, any more than the biologist need raise the question: what is life? or the physicist: what is gravity? in the context of a limited contribution to his subject. By choosing to do so Durkheim adversely affected the acceptance of a work that was undoubtedly a masterpiece.

4

Philosophies of Scientific Method: Theories of Science

For many centuries philosophers and scientists have tried to describe how science does acquire, or to prescribe how it should acquire, reliable knowledge and understanding of the world. In the last chapter we examined the chief characteristics of the scientific attitude in practice, but gave no more than superficial attention to the question of how the great body of theoretical science is thought to grow. This question has been perhaps the major concern of philosophy of science and our present purpose is therefore to consider the answers provided by the great thinkers of former times, until we come to those of major influence in the twentieth century. Some historical perspective is of great importance in studying the philosophy of science because the current position is a complex one which has arisen by embracing the wisdom of the past and at the same time by reacting against what is now seen as false or unsatisfactory.

ARISTOTLE (384–322 BC)

We saw in chapter 2 that it is to Aristotle that we owe the invention of logic, and this alone would have been sufficient to secure his immortality. But Aristotle achieved much else besides, including the formulation of both deductive and inductive methods of enquiring about natural phenomena. His deductive method was conceived as a procedure which moved by rigorous argument from self-evident premises or definitions (derived *a priori*, or independently of experience) to incontrovertible conclusions. It was through this procedure that Aristotle developed his logic, for the job of the natural philosopher was to derive a sequence of syllogisms which would lead *back* from facts or conclusions about the world until these ultimate definitions were attained. The definitions were treated as descriptions of the true nature of phenomena, and they were to be revealed by asking appropriate questions about such things as the material of which objects are made, the causes of processes, their purpose, and so on.

Unfortunately, Aristotle did not say *how* such questions were to be answered. There was no suggestion of the use of experiments, in part perhaps because the Greeks tended to regard practical activities with a certain contempt. At any rate, when it became apparent that the method of geometry could not easily be applied to other areas of knowledge – notably because basic definitions commanding universal assent were not forthcoming – Aristotle was forced to attempt the establishment of universal premises 'on the evidence of groups of particulars which offer no exceptions'.

It was the need for this inductive method – observation followed by generalization – that appears to have convinced Aristotle of the value of the empirical approach to nature. Undoubtedly his most significant contribution to scientific methodology, it was only such an empirical approach that could provide the raw material from which the 'soul' would refine the universals.

But it is symptomatic of the paradox of Aristotle that while so much of the research and description performed at his Lyceum in Athens was meticulous, revealing facts of great value over an enormously wide range of nature, a good deal else was based dogmatically on circumstantial evidence. Such however was his monumental authority, that most of his explanations of phenomena were transmitted unquestioned through 2,000 years, and it is for this reason that a minority of scholars still argue today that the influence of this unsurpassed giant of the intellect was actually an impediment to the growth of scientific knowledge, or at least to the sort of knowledge that we might now wish to characterize as scientific.

THE REFORMED INDUCTIVE METHOD

Despite its promise, the lack of clarity and precision at the heart of Aristotelian method meant that when criticism at last became possible it was delivered in severe and unremitting form. Thus it was said of Aristotle that 'when he had settled his system to his will, he twisted experience round, and made her bend to his system.' This would seem to be roughly the equivalent of the modern slogan: 'My mind is made up, don't confuse me with facts,' or at least of something along the lines: 'Give me *any* fact and I will fit it into my system.'

Francis Bacon (1561–1626)
The quotation above is from Francis Bacon, the first major philosopher

of science of the modern period. Bacon lived at a time when worship of the Ancients was just beginning to abate. But his advocacy of 'Practice' as well as 'Speculation' was still a radical philosophy, and the idea that the Greeks might not always have been right, that the present state of knowledge and conditions of life might be improved, was still daring. Bacon however was a man of the world as well as a scholar, and he handled the publication of his great work *Novum Organum* (1620) with astute diplomacy, dedicating it with much eloquence to James I. His *New Instrument* was a refined and elaborate account of the method of induction as it was to be applied to science.

The kernel of Bacon's method was, like Artistotle's, the exhaustive collection of information about the world, followed by generalization. Its ultimate pupose was the unveiling of the fundamental laws, or 'Forms', of nature for the benefit of man, and his most important single advance on Aristotle was his belief in the use of experiments designed to facilitate the discovery of facts. Moreover, he placed great emphasis on the accuracy of the information collected, saying that when there was room for doubt, the doubt should be recorded along with the information.

But, again like Aristotle, Bacon was unable to give convincing examples of his method in use. While there can be no doubt that observations and experiments are indispensable ingredients of scientific method, his attempt to derive general conclusions by induction from the mass of data was fraught with difficulties of the most fundamental kind. For example, in one extended case described by Bacon himself, he attempted to elucidate the Form of Heat. First he compiled a Table of Affirmatives, that is a list of instances in which heat is present (the sun, fire, animals and so on). Then followed a Table of Negatives, which considered the general categories of the first list, but where heat is absent (heavenly bodies such as the moon, and animals such as insects). Finally a Table of Degrees listed quantitative differences such as various kinds of flame, or animals in fever or exercise. He then outlined the procedure to be followed when these data are presented to the intellect:

> The first work therefore of true induction (as far as the discovery of Forms) is the rejection or exclusion of the several natures which are not found in some instances where the given nature (e.g. heat) is present, or are found to increase in some instances when the given nature decreases, or to decrease when the given nature increases. Then indeed after rejection and exclusion has been duly made, there will remain . . . a Form affirmative, solid and true and well defined.

Having thus 'rejected and excluded', Bacon finally came to the conclusion that 'heat is a motion,' by which we are to understand that the *cause* of heat is the movement of the smaller particles of bodies. This conception, though vague, is superficially plausible and surprisingly 'modern'. Ironically, however, it seems to have been the result of induction from insufficient data, perhaps the commonest of all the problems with inductive logic (chapter 2), for the idea of the motion of the particles of a body upon heating apparently followed from the rejection of the idea of the motion of the body as a whole. Although he certainly had the experimental techniques to demonstrate that all bodies change (i.e. expand) as a whole when heated, Bacon evidently did not notice this, so that he seems, in this instance, to have arrived at the right conclusion for the wrong reason – hardly a recommendation for his method.

To be fair, it must be said that Bacon was certainly aware in general terms of the problems of inadequate information, and it was this awareness that prompted him to collect more and more evidence before daring to proceed to the generalizing stage of his method. This of course was an unsatisfactory situation for a man so dedicated to the *Advancement of Learning,* and it was perhaps this caution which prevented him from seeing a role for the hypothesis or informed guess (with its attendant risk of error). At any rate this was the critical defect of his method, and it is a remarkable tribute to his otherwise bold and progressive outlook that he should have exerted so great an influence on the philosophy of science in later generations.

Despite his dissatisfaction with the Ancients, Bacon's inductive method, like that of Aristotle, was nevertheless profoundly influenced by the idea of the certainty of knowledge which had been derived from Greek geometry. More than two centuries after Bacon, logicians and philosophers were still under his seductive spell, none more so and with more influence, than J. S. Mill.

John Stuart Mill (1806–1873)

Mill's conception of scientific method was based on his belief that it was the purpose of science to reveal causes. By pursuing causes, he argued, we may find general laws. In chapter 2 we briefly examined some of the problems associated with the idea of causation in inductive logic and know that it is an idea that gives rise to a good many difficulties. However it is quite customary for us to think in terms of cause and effect and, loosely speaking, we understand the relation that exists between them. That is, we recognize with respect to some given effect, that if there is a change is some phenomenon which is invariably associated with or followed by a change in the effect, the phenomenon is likely to be the cause of the effect.

Procedures for unravelling causes have been proposed by a number of philosophers and the heart of classical inductivism is contained in Mill's famous Canons (or methods) of Induction, in which the influence of Bacon is obvious. The two most important Canons are stated as follows:

1 *The Canon of Agreement:* If two or more instances of the phenomenon under investigation have only one circumstance in common, the circumstance in which alone all the instances agree, is the cause (or effect) of the given phenomenon.

2 *The Canon of Difference:* If an instance in which the phenomenon under investigation occurs, and an instance in which it does not occur, have every circumstance in common save one, that one occurring only in the former; the circumstance in which alone the two instances differ, is the effect, or the cause, or an indispensable part of the cause, of the phenomenon.

Mill's Canons are, in their way, impressive rules for reasoning, but to be of value to science they must advance knowledge. At a very elementary level it is not difficult to see how they *can* work. For instance, the Canon of Agreement would seem to reveal the cause of the formation of dew in the early experiments of William Wells (see above); the one circumstance in common between the various instances of the deposition of moisture was that the wetted surface was colder than the air. Unfortunately, however, real scientific reasoning consists of a great deal more than merely noting whether certain phenomena occur together or in sequence; indeed, such a simple observation would seldom be sufficient reason for making causal connection at all. To have confidence, we should *already* have substantial theoretical knowledge concerning the phenomena, otherwise, while we might be able to draw up certain simple laws, we should make little progress towards explaining them. This problem of the priority of theory over observation leads us to examine the limitations of the inductive method in general.

Inductivism and its Problems

Thus far our conception of scientific method has been one based initially upon unprejudiced, objective observation. From his observations of particular events, the scientist derives singular statements such as 'the dewed grass was colder than the air,' or 'the mitral valve closed when the ventricles contracted.' However, science consists of knowledge about the world *in general,* not merely about isolated events (chapter 1), and this, according to the inductivist, is produced by generalizing from a finite number of singular statements to universal statements or laws. These laws would be of the form, 'cold surface gather moisture from damp air' or 'the valves of the heart keep the blood flowing in one direction.'

We saw in chapter 2 that the reliability of general laws derived by induction will depend upon both the quantity and the quality of the singular statements of evidence. For this reason the inductivist insists that many observations be made and that they be made under a wide variety of conditions. Thus in Wells' experiments, numerous different surfaces had to be examined on numerous occasions, and in Harvey's the hearts of many different species were observed. Furthermore, it must be recognized that a single well-attested observation which conflicts with the universal law (a wetted surface warmer than the air, or a heart that pumped blood in two directions at once) would be sufficient to invalidate the generalization. But provided that these conditions are satisfied, says the inductivist, the continual growth of a reliable body of scientific information is assured.

The inductivist account of scientific method does not stop with the establishment of universal laws; these laws are to be used as a basis for the next stage, for science seeks explanations of the way the world works and wishes to make predictions of future events. It is in this connection that the inductivist is happy to acknowledge the role of *deduction* in his method, because he will want to deduce predictions and explanations from his universal laws and these could be derived by means of old-fashioned syllogistic reasoning. For example, having made innumerable observations on various gases under a wide variety of conditions of pressure and volume, he will have drawn up the universal law that (at constant temperature) 'the volume of a gas is inversely proportional to its pressure'. This will serve as the first premise of his syllogism, the others being:

Air is gas
therefore if the pressure of air is doubled the volume will be halved.

The truth of the first two premises is ascertained by experience; if they *are* true, the conclusion (prediction) must be true. But the truth of the deduced conclusion rests entirely on the assumption that the processes which led to the first two premises, namely observation and induction, can indeed be justified.

Such justification is, in the strict sense, hard to come by. So much so that most modern philosophers think that inductivism is fatally flawed as a methodology on this account. We may briefly consider the three most important arguments. The first is that the inductivist's claim that observation can be a sure basis for scientific knowledge is not borne out by closer examination. This idea assumes, with respect to vision (similar arguments apply with the other senses), that the human eye works like a camera, and that the observer 'sees' whatever is projected on to the retina. Unfortunately, this is far too simple a model, as we all know from those 'trick' pictures (see figure 3, p. 64) in which the 'mind' decides to see either a vase or two human profiles, a rabbit or a bird, an old woman or a young girl. The point is that what one 'sees' (or hears, feels, smells etc.) depends only in part on the immediate sensory experience, but also in part on accumulated past experience and on expectation. Observation is not, then, 'objective' in the sense that several observers viewing the same object necessarily have the same experience.

The second argument against the inductivist case is perhaps no more than an extension of the first. It is that several observers viewing one object will formulate different singular (observational) *statements* about their experience. Even if we allow that their individual perceptual experiences might conceivably be direct and independent of past experience, their public statements cannot be expressed without reference to prior theoretical understanding. Even the simple singular statements given above presuppose some elementary theory. For example, the word 'colder' implies an awareness that bodies can be at different temperatures, and the phrase 'the ventricule contracted' already seems to assume a knowledge of the heart as an organ that moves the blood. The importance of this is that it shows not only that observations are *preceded* by theories (i.e. they do not *lead* to theories as the inductivist claims), but also that the reliability of the observations will inevitably be determined by the reliability of the theories upon which they depend. In order to verify a singular observation statement we are forced to appeal to our theoretical knowledge, not to more basic observation statements. But our theories are fallible, and our observation statements cannot therefore be an infallible basis for scientific knowledge.

The third argument against inductivism is the logical problem of induction itself, which we encountered in chapter 2. If we accept the first two arguments, this one might be thought irrelevant, since even if inductive generalization could be proved sound, its conclusions would be no more secure than its initial observational data. But the problem of induction has not been resolved. There is no rule which can tell us when (if ever) we have collected sufficient observation statements to justify generalization, and we cannot know what *kind* or variety of observations to make without prior theoretical knowledge of the problem. Equally, we cannot know whether a single observation that appears to conflict with a universal law actually *counts* as doing so without a theoretical background. For example, would a heart with faulty valves refute the law that the valves ensure the unidirectional flow of blood?

The impossibility of justifying inductive generalization would be countered by the more sophisticated type of inductivist in terms of some sort of probability theory. Even if we have no absolutely secure basis for scientific knowledge, he would argue, we can at least be sure that cautious generalization will give us knowledge that is *probably* true. This is the best that we can hope for, and it is to be achieved by induction.

This view is likely to attract many more adherents than the more extreme, naive, claim of a wholly objective science. Yet, according to many philosophers, even this milder form of inductivism, does not hold. This is because any universal generalization makes (by definition) predictions about an *infinite* number of possible future situations, yet has been derived from a *finite* number of observation statements. The probability of the truth of the generalization is found by dividing the finite number by infinity and it is therefore always zero.

This theoretical point notwithstanding, the weaker inductivist position does have powerful intuitive appeal; it seems 'obvious' that there must be something in it. Perhaps this is simply because we do customarily think inductively. It *does* seem more probable that a person who routinely drinks heavily before driving will have an accident than someone who is always sober; and part of the reason for this is that we have a good deal of knowledge of the *mechanisms* by which alcohol enhances the sense of confidence, yet slows the body's reflex reactions. For much the same reason we do expect the sun to set this evening, not just because it has always done so in the past, but because we trust the laws of astronomy. Although we know that in reality all inductively derived knowledge is founded on uncertainty, many people will agree with Hume that our pyschological make-up compels us to behave as if induction were a logically valid procedure.

THE HYPOTHETICO-DEDUCTIVE METHOD

Whatever one's attitude to inductivism, there seems to be no doubt that the prodigious advance in the growth of knowledge that has come in more recent times is the result, not just of the exhaustive collection of new factual material and of generalizations from the facts, but rather of a radically new approach. This is widely identified as the 'method of hypothesis,' because this collection and analysis of information is here guided by a preconceived idea.

William Whewell (1794–1866)
The move started somewhat tentatively in the nineteenth century in the hands of the versatile Cambridge Divine, William Whewell, a contemporary as well as an adversary of J. S. Mill. In his monumental work

on *The History and Philosophy of the Inductive Sciences* Whewell examined in great detail the role played by observation and experiment and, most significantly, of ideas or 'happy guesses', in the advance of scientific knowledge. Observing with generosity that in Bacon's time it was possible only to theorize upon how science ought to proceed, he noted that nineteenth-century thinkers enjoyed the great advantage of being able to look at how it had, in fact, proceeded. There are, he said no universally applicable methods of induction. Bacon's collection of data could lead to sound theory only 'by means of a true and appropriate conception', by which he meant that a hypotheses must be invented at an early stage to account for the observed facts. This was an advance in understanding of profound importance. Not surprisingly, Whewell could not explain by what means the hypotheses were arrived at, but his own words are of considerable interest in revealing his obvious appreciation of the elusive faculty of creativity:

> The conceptions by which facts are bound together, are suggested by the sagacity of discoverers. This sagacity cannot be taught. It commonly succeeds by guessing; and this success seems to consist in framing several tentative hypotheses and selecting the right one. But a supply of appropriate hypotheses cannot be constructed by rule, nor without inventive talent.

Whewell's ideas about scientific methodology marked a fundamental divergence from those of his predecessors, most notably in his rejection of the view that 'truth' has only to be revealed by systematic search for the laws of nature, and his substitution of the Kantian idea that it is rather to be found in the mind of the investigator. This truth takes the form of descriptions or explanations that are invented by the mind to account for observed phenomena.

Karl Popper and the Method of Falsification

The most influential advocate of this conception of scientific methodology has been Karl Popper. His views were first published in 1934 in *The Logic of Scientific Discovery* (translated into English, 1959) but the title of his more recent book *Conjectures and Refutations* (1963), encapsulates the essentials of his 'method of falsification'. Hypotheses are to be developed and attempts made to falsify them through empirical research. In Popper's own words: '. . . there is no more rational procedure than the method of trial and error – of conjecture and refutation; of boldy proposing theories; of trying our best to show that these are erroneous; and of accepting them tentatively if our critical efforts are unsuccessful.'

Popper grew up in Vienna, and matured intellectually during the time when the so-called Vienna Circle of philosophers was propagating its doctrine known as logical positivism. Popper had, however, very little direct contact with this group, although their views on the status of scientific knowledge and in particular on the criterion by which it was to be distinguished from nonsense, did have a significant influence. The Circle's famous verifiability principle claimed that the only meaningful propositions were those either of logic and mathematics (which, if true, were tautologies) or of empirical science. The meaning of scientific propositions could, it was claimed, be verified by observation and the principle was used to attack theology and metaphysics, not merely as non-scientific, but also as non-sensical, since their propositions clearly could not be so verified. Unfortunately for the logical positivists, the principle of verifiability proved to be unworkable as a criterion of demarcation for it excluded all the general propositions of science which cannot be verified on account of the problem of induction, and in many other connections was shown to raise more problems than it resolved.

At any rate, Popper's basic and most fertile insight at this stage was that while even scientific hypotheses could *not* be verified, they could nevertheless be shown to be false. (It was in this sense that they differed from metaphysical propositions – see below.) Because this view has been so influential, and because for many it still epitomizes *the* modern scientific method, we must examine its logical foundations.

In the hypothetico-deductive method it is sufficient that we take as our starting point the definition of an hypothesis as 'any statement that is used as a premise, logical implications of which may be tested by comparison with facts ascertained by observation'. Although in modern terminology the method is described as hypothetico-deductive, it is important that even this process of testing scientific hypotheses be seen, in the broadest sense, as inductive. By this we mean that when an hypothesis is tested and accepted (i.e. not falsified), the evidence for the acceptance is not deductively conclusive in the way of a (tautologous) mathematical argument, such as:

$$x^2 - y^2 = (x+y)(x-y).$$

Rather, the evidence merely affords more or less strong inductive confirmation. The status of scientific knowledge is thus lower, in the logical sense, than deductive knowledge. The 'naive' falsificationist grants that we are never able to say that a scientific theory is absolutely true, yet we can say of some well-tested theories that they have a great deal of support. For practical purposes, this is borne out for many theories in their daily applications.

The chief value in the use of the term 'hypothetico-deductive' is that it clarifies the important distinction between the separate acts of discovery and of justification. It is the former process which is akin to Whewell's 'sagacity that cannot be taught' and which is characterized by such words as inspiration, imagination and intuition. It is the process of justification on the other hand, which is the unique safeguard of scientific method, eliminating the wilder fantasies of the imagination and subjecting all new ideas to a dispassionate, sceptical analysis. In this process, furthermore, science itself may most clearly be distinguished from other creative processes. By deducing consequences from the hypothesis and by comparing these with empirical data, the hypothesis may either be falsified or it may receive support. If it is falsified it must be abandoned; if it is confirmed it survives to fight another day.

As we have said, Popper's singular contribution has been his sharp distinction between the attempt to prove and the attempt to disprove scientific statements. Although no number of confirmatory observations can permit us logically to verify the universal statement 'all birds can fly' (because of the empirical and logical problems of induction), a single observation of a flightless bird would permit us to conclude that it is *not* the case that all birds can fly. The attempt to prove theories true is futile because it is logically impossible. What is possible is to deduce the falsity of theories from singular disconfirmatory statements.

Here it should be briefly noted that there is a distinction between falsification at the level of logic and that at the level of method. Given that 'all ostriches are birds', discovery of a single ostrich *logically* proves false the assertion that 'all birds can fly'. However, we might insist that this species is not a bird at all, but a feathered reptile or, indeed, that it cannot be a bird *because* it is flightless. There is nothing, in fact, to stop us from rejecting all falsifying evidence whatever, such that methodological falsification, like logical verification, is impossible (see the next section). Popper resolves this dilemma essentially by an appeal to good common sense, for if we did seek the impossible and yet continue to reinterpret any evidence threating our hypothesis, we should quickly make a nonsense of scientific method. We must, therefore, as an article of method, set out our hypotheses as clearly and unambiguously as possible, so that they may be rigorously tested by attempts at refutation; a clear and precise hypothesis will be more readily falsifiable than a vague one.

This is an important point for Popper because it leads him to his own criterion of demarcation between science and non- or pseudo-science. To be scientific, a hypothesis must be logically falsifiable. That is, there must be – at least in principle – some sort of conceivable observation statement that would contradict the hypothesis. Harvey's hypothesis concerning the movement of the blood '. . . as it were, in a circle' is readily falsifiable in this sense. So is Torricelli's that the height of a vertical column of mercury (such as in a barometer) is proportional to the 'weight of the sea of air' supporting it. Statements which cannot be falsified and which consequently, according to this viewpoint, have no informative content, would include those mathematical propositions and definitions which are tautologies (e.g. 'the angles of an equilateral triangle are all 60°'); vague assertions such as occur in popular horoscopes ('the middle of the month could be good for business transactions'); and normative statements which assert not what is, but what it is considered ought to be ('low direct taxation is desirable').

This distinction inevitably leads the falsificationist into sensitive areas. Popper himself, for example, has claimed that Freud's psychoanalysis is not a science precisely because its theories can explain everything that an individual can possibly do or experience. The point here is that a genuine scientific theory does just the opposite: by making only limited claims about the world it actually excludes most of what could *possibly* occur, and is itself excluded if what it excludes occurs. A similar situation obtains with Marx's theory of history, although critics might argue that it is less a question of unfalsifiability than of its having been falsified and yet retained. Its adherents, they say, cling on to the theory by constant (methodological) modification, so that for them at least it has not been falsified.

Leaving this particular controversy aside, we can at any rate say that for the falsificationist a theory qualifies as part of the body of scientific knowledge by being falsifiable but not yet falsified. To say of a theory that it is falsifiable is to say that it has informative content, and the more informative it is the more falsifiable it must be. The more falsifiable a theory is the better, so it is the scientist's job to advance 'bold conjectures' in preference to cautious ones. In Popper's sense, the boldness of a proposition is a measure of its generality, for a more general proposition is bound to offer more opportunities for falsification than a more limited one that it subsumes. For example the general law stating the inverse relation between pressure and volume of any gas is more falsifiable than a specific law which refers only to air. If experiments with air show the relation to be other than inverse, they will have falsified the specific law *and* the general one. The reverse however is not true, because experiments with oxygen or nitrogen alone, which also falsified the general law, would leave the specific one untouched.

The act of falsifying a theory is, for Popper, a high point in science. Indeed, it is the moment at which its body of knowledge grows. The concept of growth and progress is a crucial one. Whereas for the inductivist science progresses by the accumulation of more and more observations and by the cautious induction of theories from the observations, for the falsificationist all observations are themselves inevitably preceded by theories (they are 'theory-laden'), and progress therefore occurs by making bold speculations which can account for more observations and survive the tests that falsified earlier theories. It is not enough that a new theory be falsifiable. For science to grow, it must be *more* falsifiable than the one it replaces, for then it will be more general and more informative.

Unfortunately, it is by no means always a simple matter to decide whether a theory *has* been falsified. There are two reasons for this. First, there is the problem that any hypothesis under test as a candidate for the body of scientific knowledge is bound to rely upon one or more additional *auxiliary* hypotheses which have already been tested independently and have achieved that status (see chapter 3). Despite their wide acceptance, these auxiliary hypotheses are not 'certain' (they have merely not been falsified) and in attempting to determine the logical status of the new hypothesis it can never be known for sure whether it, or an auxiliary hypothesis, is confirmed or refuted.

The importance of the auxiliary hypothesis can be readily illustrated in connection with Robert Boyle's work on gases. His 'law' that the product of pressure and volume is a constant invites an infinite number of empirical predictions which, if borne out, give support to the hypothesis and, if not, refute it. In the strict sense however this is an over-simplification, for it depends upon the reliability of the auxiliary hypothesis (Torricelli's) about the pressure exerted by the atmosphere. If this pressure is not added to the pressure experimentally exerted on the enclosed air by mercury, the product of pressure and volume would not be constant. Table 1 demonstrates the point using some of Boyle's own results (simplified).

Table 1   *The concept of atmospheric pressure as an auxiliary hypothesis in Boyle's Law*

Experimental pressure exerted (P_1)	Experimental pressure plus atmospheric pressure (P_2)	Volume of air (V)	Product (P_1V)	Product (P_2V)
0	29.1	48	0	1397
10.1	39.2	36	364	1411
29.7	58.8	24	713	1411
48.8	77.9	18	878	1402
88.4	117.5	12	1061	1410

In practice the falsificationist will tend to assume that any auxiliary assumptions are 'true' and it is then possible for him to test by logic the status of his new hypothesis. If, for example, a prediction deduced from the hypothesis turns out to be false (P_2V in table 1 is not constant) then the hypothesis itself must be taken as false, as (effectively) the sole premise in the argument. But it may happen that the predictions deduced from a given hypothesis could, in fact, be equally well derived from one or more alternative and competing hypotheses. The problem of choosing between them is not always easy. Although it has no strict justification in logic, a 'principle of simplicity' is commonly advanced by scientists. The idea is attractive to the falsificationist because the simpler a hypothesis is the more falsifiable it is likely to be. For example, if we test the relation between volume and pressure of a gas and find that we can express the results graphically as a straight line of negative slope, we should be likely to conclude that this confirms the hypothesis that volume and pressure are inversely proportional. However, without performing an infinite number of experiments it is always possible that the individual points on the graph (which do show *some* variations from the linear) should in reality be joined by a line of bizarre complexity. The latter hypothesis would be difficult, if not impossible, to falsify, and the former would certainly be accepted. Rival hypotheses of apparently equivalent simplicity would ideally be separated by means of some form of crucial test which would compare different, and preferably incompatible, observational predictions. Although there are logical problems even with a situation of this type, for practical purposes we should then be inclined to accept the hypothesis which received most satisfactory support.

The second problem in deciding when a theory has been falsified is again essentially that of distinguishing between falsification at the level of logic and that at the level of methodology. When a theory has already gained substantial confirmation from other predictions, how are we to decide if the failure of a new test is sufficient to falsify it? A famous example was the early discovery that the orbital path followed by the planet Uranus was not that predicted by Isaac Newton's all-embracing equations. Newton's work had received enormous support elsewhere and, instead of attempting to overthrow his theory as a result of the awkward new observtion, it was boldly predicted that an as yet unknown planet was responsible for the eccentric movement of Uranus. Subsequent searches led to the dramatic discovery of Neptune and, in the new situation, Newton's theories accurately predicted the observed path of Uranus and gained significant further support.

It is not just with the wisdom of hindsight that we say this procedure was a justified defence of Newton. Although it may at first appear to be a device, similar (according to Popper) to that used by some Marxists, which is merely designed to 'save' the beloved theory, the important difference is that the re-formulation of Newton's equations in the light of the predicted new planet, could be *tested* independently. It *was* tested and found to hold. The orbit of Uranus could, after all, be predicted by Newton's theory, though in a form now somewhat different from that set out by Newton himself.

If the problem had been 'explained away' (as distinct from genuinely explained), let us say by postulating some occult force in Uranus itself which caused its unpredictable path, then no tests would have been forthcoming. This *would* be a mere device, one which is called by falsificationists an *ad hoc* hypothesis, and whose sole purpose is to evade refutation of the main theory. This distinction is important and we might once again refer to the work on the pressure of the atmosphere to make it clear. The Frenchman Blaise Pascal tested Torricelli's idea about the weight of the sea of air by carrying a barometer to the summit of the Puy de Dome in central France. He was at the same time concerned to disprove the Aristotelian notion that mercury remained in the upright tube because of nature's alleged 'abhorrence of a vacuum'. The results refuted the latter hypothesis, because the height of the mercury fell as the mountain was climbed. However, some of those most committed to it merely introduced the *ad hoc* hypothesis that the strength of nature's abhorrence, rather than the pressure of the atmosphere, is related inversely to altitude. It would always be the case that an idea of this sort would be true, so that for science it has no interest.

The moral in this, according to the falsificationist, is of course that the scientist should never resort merely to *ad hoc* defences, and the best way to ensure against this is for him actually to *welcome* the falsification of his most cherished theory as evidence of something far more important – the growth of scientific knowledge.

But what should be the response to the confirmation of a hypothesis? When a prediction derived from a new hypothesis turns out to be true, this provides no evidence in the strict deductive sense for the truth of the hypothesis. Yet it has to be granted that the status of the hypothesis in this situation does appear to be different from that following a prediction that is false. The hypothesis has certainly not been proved but it has gained some support. It is only for the most extreme falsificationist that the refutation of hypothesis is the *only* way in which science grows. For most the confirmation of hypotheses is also important.

Here it is useful again to distinguish between bold and cautious hypotheses. With bold hypotheses, confirmation will be *more* important than falsification. A bold hypothesis ranks highly is information and in predictive power and when confirmed adds to science something that previously was thought unlikely or was not thought of at all. Such was the case with the discovery of the planet Neptune or with Harvey's discovery of the circulation of the blood (especially when the anatomical route for the passage from arteries to veins was revealed by Malpighi). But the falsification of a bold hypothesis has little importance for science as a whole. It is no more than the fate that should be anticipated for most implausible ideas.

With cautious hypotheses, falsification is more significant than confirmation. For instance, the claim made by an Aristotelian physicist (or, for that matter, by many modern non-physicists) that an object can move only as the result of a force acting on it, was certainly thought to be free of intellectual risk. Since Newton however, physicists have regarded it as false. Because it was based upon premises thought to be self-evident (i.e. 'true'), its refutation is particularly informative. But the mere confirmation of a cautious hypothesis is trivial because it represents still more support for something already well known.

Strengths and Limitations of Falsificationism

We have seen that the falsificationist overcomes the inductivist's problem of the 'theory-laden' nature of observations by acknowledging that theory precedes observation. Then, switching emphasis from the verification of theories that are 'drawn out' of his observations to the refutation of theories that are 'thought up' to account for observations, he also claims to have circumvented the purely logical problem of induction itself.

We have, however, seen that the claim that theories *can* be falsified is itself not without problems. While the discovery of an ostrich necessarily falsifies the simple claim that 'all birds can fly' at the level of logic, when we get into real science the situation is immeasurably more complex because every assertion made is bound to rest on numerous auxiliary hypotheses. If a prediction made on the basis of a real scientific theory proves false, we can never be sure, even logically, whether the theory, or some auxiliary assumption, is false.

There is however still another weakness at the heart of falsificationism. It argues that there is an essential qualitative distinction between the attempt to verify a theory, which can never be decisive, and the attempt to falsify, which can. However, the case against inductivism – that observation statements are theory-laden – can be turned also against the falsificationist in much the same way. That is, if an observational prediction of a theory indicates that the theory has been falsified, we can be sure that this is so only if the observation statement itself is reliable. But wholly reliable observation statements are not available (or, at least, we can never know that they are) so that we have no way of telling if it is the theory, or the observation statement, that is false. Thus the falsification of theories is, strictly speaking, no more certain than verification.

For many a practising scientist, the claim that we can neither prove nor disprove theories is deeply uncongenial. The scientist 'believes' in the reliability of his science and may well be tempted to conclude that philosophy of science, just like much of philosophy in general, is merely trying to tie him up in knots by showing that what seems obviously to be the case cannot possibly be so. The reader who is on the brink of exasperation may feel relieved to learn that one of the most powerful arguments against both inductivism and falsificationism is not subtly philosophical but derived directly from the history of science itself. Although inductivists and falsificationists customarily draw upon examples from the past (as, indeed, they must, if their accounts purport to describe what has actually occurred) it happens that the *same* examples (e.g. Darwin's evolutionary theory) are sometimes used to illustrate their *incompatible* theories. It also happens that other examples can be fitted to only one theory (e.g. Mendel's genetics to inductivism; Harvey's physiology to falsificationism) or to neither (some interpretations of the 'Copernican Revolution'). This casts the gravest doubt upon any claim that there is such a thing as *the* scientific method. Perhaps it is rather that different scientists (or sciences) use different methods at the same time or at different periods in history.

Certainly there has been a move during the last twenty years to get away from the idea of science as a reciprocating interaction between individual theories and limited empirical data. It is the highest complement to the influence of Karl Popper to say that the most important formulations in philosophy of science during recent years have been to a substantial degree a reaction against his work, just as his own was a rejection of inductivism and the verifiability principle of the Vienna Circle.

Careful reading of the history of individual sciences reveals that their growth has been not bit-by-bit as the falsificationists and inductivists would have us believe, but rather as structured 'wholes' with a distinct organic continuity through time. Thus, for example, the development of astronomy with the growth of the Copernican system shows it to have unfolded in a programmatic fashion over a period of many decades. Much the same can be seen in physics with the work leading up to Boyle's law relating the pressure and volume of a gas; in biology with the development up to and beyond Darwin's *Origin of Species*; or in chemistry with the fluctuating fortunes of the phlogiston theory and the eventual discovery of oxygen. Any notion of what science is *really* like must account for these protracted periods of growth which give every appearance of having been 'guided' by some unity that is altogether bigger than the individual hypotheses and observations which are embraced. We shall consider the two most influential theories of this general kind.

Thomas Kuhn and the Scientific Paradigm

One of the first attempts to give to science the organic character that seems to be demanded by its history was that argued by Thomas Kuhn in his enormously influential book *The Structure of Scientific Revolutions* (1962). Kuhn's view of science thinks in terms of communities of scientists rather than of individuals. To this extent it is a sociological theory. Despite the title of his book, the most characteristic feature of the scientific enterprise as depicted by Kuhn is its conservatism, which is seen as the consequence of the prolonged 'indoctrination' that scientists receive during their training. This is an indoctrination within the confines of what Kuhn calls a 'paradigm'. In the dictionary the word paradigm is said to mean 'model' or 'example', but in Kuhn's sense it means far more than this: a great research tradition, a whole way of thinking and acting within a given field. General examples would be Newtonian mechanics in physics, the concept of the atom in chemistry and that of evolution in biology. More specific examples are the views as to the position of the sun and planets held by astronomers before and after Copernicus, or the ideas on inheritance of characteristics that sprang from the work of Mendel.

According to Kuhn it is the paradigm which represents the structured whole of a given science and which also guides the research activities of its community. This is because the paradigm represents the totality of the background information, the laws and theories which are taught to the aspiring scientist *as if* they were true, and which must be accepted by him if he in turn is to be accepted into the scientific community. The work of the community is likened by Kuhn to 'puzzle-solving' and the sum of this activity constitutes his 'normal science'. Normal science, working within the paradigm and not questioning its authority, is a highly cumulative process. In this limited sense Kuhn's notion resembles that of the inductivists, although more fundamentally he is opposed to that position in his implied assertion that all observations are theory-laden (because they are determined by the paradigm). We shall see shortly that his overall view of science departs radically from that of either the inductivists or the falsificationists.

Not only is normal science cumulative. It is also stable and successful – within its own terms. Stability and 'success' are to be seen as limiting fuctions of the paradigm, for the latter exerts its control by ensuring that normal science tackles only problems which it has every expectation of solving. (This is neatly encapsulated in P. B. Medawar's famous description of science as *The Art of the Soluble*.) It is in this sense that Kuhn likens it specifically to the attempt to complete a jigsaw or crossword puzzle. There *is* a solution and a failure to find it reflects badly, not on the paradigm, but on the scientist himself. It is because of this cautious and selective approach that we have the impression that normal science progresses rapidly by comparison with any discipline willing to wrestle with problems that arouse controversy and passion.

This also means that the major concern of normal science is not the search for new phenomena or for substantive novelties, but the refinement of the paradigm which, of course, is never perfect. Kuhn describes such 'mopping up' work as one of the chief (and unflattering) characteristics of normal science. It tends to promote specialized research by esoteric groups whose primary motivation is to uncover facts which are 'significant' in terms of the paradigm, which conform to its theoretical predictions, and hence provide confirmation. Thus the 'normal' scientist is not concerned with the refutation of theories and, indeed, Kuhn explicitly rejects falsificationism as a methodology. Popper's answer to this attack is to acknowledge the existence of normal science, but to reject it in turn as merely bad science. From Popper's viewpoint it is perhaps that normal science is performed by those whom Victor Hugo scathingly characterized as the 'intellectual proletariat of research' – individuals who do science for a living rather than as a vocation, who are insufficiently critical and poorly trained.

Be this as it may, it is because normal science takes its paradigms for granted that there is a tendency to rigidity, and to the rejection as metaphysical of problems that appear too difficult to solve. According to Kuhn, the stability given to a discipline by the acceptance of a paradigm is a reflection of that discipline's maturity. The paradigm is, in a word, the criterion of demarcation between science and non-science. Thus sociology, says Kuhn, has an enormous quantity of empirical data, yet for the most part it is too weak and confused in its theories (which neither explain nor predict with significant success) to have developed a discernible paradigmatic structure to which all sociologists can subscribe. Physics, on the other hand, has a number of highly articulated theoretical traditions which largely account for its observational data and give notable security to its practitioners.

That changes of paradigms occur at all may seem surprising in view of the sophisticated protectionist attitude within which normal science operates. Yet it is to the history of science that Kuhn appeals in arguing his theory and it is the revolutionary paradigm shifts which he identifies as the most important source of scientific 'advancement'. He develops in some detail the example of changes within the science of optics. Under the influence of Newton it was held that light consisted of material corpuscles. Within this framework it was natural that research effort should be expended on the attempt to measure the pressure exerted by the corpuscles on solid objects. Such evidence was not forthcoming, and by the nineteenth century the paradigm had changed, light being considered as a wave motion. It was now appropriate, not to look for pressure effects, but rather to investigate phenomena such as reflection, refraction and interference. Even this paradigm was not wholly satisfactory and present-day textbooks (which, above all, embrace and articulate *prevailing* paradigms) describe light as photons which, ironically, share some of the characteristics of particles and some of waves.

In the mature sciences psychological commitment by a community to its paradigms is enormously strong and it is for this very reason that events leading to the fall of a paradigm, and to its replacement by one that is more comprehensive, are so traumatic, in much the same way as are political upheavals. New discovery, according to Kuhn, begins with the awareness of anomaly – that is that nature has in some way violated the expectations aroused by the paradigm. Because of the nature of normal science, the start of a revolutionary phase is at first resisted. The significance of the observed anomaly may be missed altogether, or at least it may be dismissed. The paradox associated with paradigm-shift is that genuine anomaly is acknowledged only when there *is* a detailed body of expectations from which it sticks out like the proverbial sore thumb. Without the 'received' background pattern, the unexpected result would not be seen.

The crisis resulting from new and unexplained discovery may, of course, be large or small. In most cases normal science will ultimately confront the phenomenon and prove cabable of coping with it. An example would be the deviation in the orbit of Uranus which was mentioned above. In others, as perhaps with so-called unidentified flying objects, no universal solution will be advanced and yet the explanations from outside the prevailing paradigm will be seen as having insufficient plausibility or testability to precipitate a major re-think. Accordingly, the problem will be set aside to await further data. In the third case (for example, the overthrow of Ptolemy by Copernicus), where the new theory appears to represent a fundamental improvement, there is typically a period of extraordinary science which may appear random, or almost desperate. The advocates of the old paradigm and the new may line up on opposite sides. Even in a highly quantified

Figure 3 *An example of 'visual gestalt': is this a picture of an old woman or a young girl? Or both?*

and precise science such as astronomy, research during a period of genuine crisis may resemble that in the less mature sciences, in which no clearly accepted paradigms have yet emerged. Thus there arise competing 'schools' of thought whose contrasting interpretations of observational data lead Kuhn to describe them as 'living in different worlds'.

The experience of a scientific revolution is to be likened to that of 'religious conversion', and Kuhn uses the example of the visual gestalt in which the raw data, say lines on paper, do not change in the minutest detail, and yet the mind, before and after the experience, interprets them in radically different ways (see figure 3). Minor instances of the psychological nature of this experience are actually rather common in science even on a day-to-day basis and will be familiar to anyone who has suddenly realized what it is like at last to 'see' the object of his search on a microscope slide. Major cases are dramatic and absolute because the scientist's strenuously earned commitment to his paradigm is broken only by a pronounced failure, coupled with the *simultaneous* transfer of allegiance to what is, in effect, a new 'world view'.

After the break, individuals are responding to the same 'world' as if it were quite different, and in this sense again Kuhn's comparison with political revolution is apt. The revolution brings changes of a type so radical that they would have been prohibited by the old paradigm (institutions). The choice between old and new is therefore that between incompatible ways of life and it is hardly surprising that paradigm shifts have almost always been wrought at the hands of the young, or at least of those new to a particular field. Conversely, allegiance to the old paradigm has not uncommonly ceased only with the death of its advocates. As Max Planck put it: 'A new scientific truth is not usually presented in a way that convinces its opponents; rather they gradually die off, and a rising generation is familiarised with the truth from the start.'

Here we reach what some would regard as the most fundamental contrast between Popper and Kuhn. The latter is saying that paradigms are *not* rejected by falsifying comparisons with nature, but only after comparison with nature *and* with an alternative paradigm. The scientist is simply too committed to his present paradigm ever to reject it on account of anomaly, for this would imply the rejection of science itself, and research would thereafter be impossible. Thus 'the act of judgement that leads scientists to reject a pardigm is always simultaneous with that which accepts another.' For Kuhn the gestalt shift is fundamentally a psychological process which is 'non-rational'; no strictly logical argument can account for the change nor, ultimately, defend the one paradigm against the other. It is not of course that logical arguments have no part to play, but rather that the revolutionary crisis airs a variety of fundamental standards and metaphysical assumptions that *can* be resolved – at the time of the crisis at any rate – only in terms of intuitive judgement.

While there is no *a priori* reason by which a new paradigm may be judged superior to its predecessor, Kuhn is at pains to point out that he is 'a convinced believer in scientific progress'. This is in spite of the irrational element at the core of his theory. Kuhn argues that progress occurs in normal science as each new puzzle is solved, as well as through revolutions as the new paradigm promotes more successful puzzle-solving by asking new questions and suggesting different observations and experimental techniques. At least in retrospect, we say that science was advanced by Einstein's revolution in physics because the puzzles solved by Newton were solved also be Einstein, together with a small number of critical puzzles upon which the Newtonian paradigm pronounced wrongly or had nothing to say. Few people would assert that Newton was *wrong,* yet in some crucial respects Einstein's theory seems to be 'better'.

Many would then be inclined to go one stage further than this and say that while Einstein's paradigm is perhaps not the final truth, it is closer to the truth than Newton's. If this cannot be claimed, they would argue, it is hard to see in what sense the expression 'closer to the truth' could have meaning. But it is about this very idea of 'truth' that Kuhn himself does feel uncertain. He wishes to suggest that there may be an analogy between the progression of scientific ideas and the evolution of living organisms. There *is* progress in science, but it has occurred 'without benefit of a set goal, a permanent fixed scientific truth, of which each stage in the development of scientific knowledge is a better exemplar'. For Kuhn, there is no evolution *towards* anything.

Imre Lakatos and Research Programmes

The second theory which derives from history a view of science as consisting of structured, organic wholes, was presented by Imre Lakatos in 1970. In his 'sophisticated methodological falsificationism, he speaks of sequences of theories which are welded together into continuous 'research programmes'. The programmes offer the scientist guidance, both as to which research problems to avoid and which to pursue. The former constitutes the 'negative heuristic' of the programme. This is characterized by a 'hard core' of background information which, says Lakatos 'is irrefutable by the methodological decision of its protagonists', and has emerged over a long period of trial and error. Once established, the hard core is sacrosanct, and at all times is defended against attack by a 'protective belt' of auxiliary hypotheses and observations, the function of which is to be tested and refined, or completely rejected, while protecting the core. Thus in a given research programme various auxiliary hypotheses will be added to accommodate new anomalies, and the programme is said to be progressive so long as the range of empirical observations that it accounts for grows, particularly in the sense of its success at predicting novel facts. The programme is described as degenerating if it cannot do this.

As an example of a progressive research programme, Lakatos takes William Prout's daring theory (1815) that the atomic weights of all elements must be whole numbers. This, he says, was at first submerged in an ocean of anomalies. The hard core of the programme was that all atoms are compounded of hydrogen atoms, (which had already been given the atomic weight of 1), and the champions of Prout's programme, having once decided to accept the hard core, could explain anomalies only in terms of the inadequacy of the protective belt. Thus when there was set against them the apparently falsifying observation that the atomic weight of chlorine is 35.5, the defence had to be in terms of the question: 'But was the chlorine a pure chemical element?' Since it was, according to the best evidence of the day, the Proutians had then to overthrow the theoretical base of this evidence by revolutionizing the techniques of analytical chemistry. Although for many years the theory was considered dead, new ideas on the structure of the atom emerged during the early years of the twentieth century and, in the event, a new version of Prout's hypothesis was established. The Proutian's guide for their radical and persistent support was provided by the 'positive heuristic' of the research programme, a machinery, that is, for solving problems, digesting anomalies and so on. This it was that showed them which particular problems to pursue in their protracted struggle to develop the necessary 'refutable variants' of the protective belt.

Lakatos discusses as a particular example of the positive heuristic Newton's development of his gravitational theory of a planetary system. Newton arrived at his inverse square law by considering the simple model of a point-like planet moving elliptically around a point-like stationary sun. However this model conflicted with his own third law of motion (part of his hard core) and it had to be replaced by that of the sun and planets revolving about their common centre of gravity. This again had to be made to apply to more nearly real situations in which the sphere-like planets were attracted, not only by the sun, but also by each other. This progressive development of the programme involved great mathematical problems and, according to Lakatos, Newton resolved them under the guiding influence of his positive heuristic rather, that is, than responding to the many anomalies that he knew well to exist. It is in this important respect that the positive heuristic is said by Lakatos to protect the scientist from the confusing influence of counter-observations; only by means of its defensive shield is the early development of his research programme made possible in relative isolation from potential falsifying instances. The programme can then become established on a strong footing before the latter have to be faced. This overcomes the awkward historical fact that many theories which, with hindsight, grew to be regarded as of signal importance to science would, in falsificationist terms, have been strangled at birth on account of unexplained anomalies.

In the case of Newton's planetary system, examination of observational data showed that some of the anomalies could be explained by this model, although many could not. The unexplained anomalies prompted Newton to further theoretical developments, for instance to consider bulging, non-spherical planets. The positive heuristic also suggested a practical research policy, for example on improving tele-scopic techniques and on sensing gravitation forces in the laboratory. Technical and experimental development would, it was believed, eventually refute the counter-evidence and provide confirmatory observations, and there was never a more dramatic example of the latter than when the planet Neptune was first predicted and then observed.

Lakatos agreed closely with Popper about the steps that may legitimately be taken to 'save' a research programme. The protective belt may be modified in order to defend the hard core in any way, provided that it is not merely *ad hoc* and, accordingly, untestable. For a programme to be successful and, indeed, *scientific*, means in fact that it must maintain its own ordered and cohesive framework to guide research, and it must continue to generate novel phenomena which can be tested. For Lakatos therefore, as for Popper, sociology and psychoanalysis cannot be regarded as sciences because they fail to meet the first and second criteria respectively.

Lakatos's theory was formulated, he says, as an attempt to develop and improve through 'Popperian spectacles' on the falsificationist account, in particular by giving its theories continuity through time. In achieving this the theory also revealed its debt to Kuhn. Another similarity with Kuhn's theory is the problem of determining which of two competing research programmes is 'better' or at least is to be preferred. To say that a progressive programme is to be preferred to a degenerating one is to beg the question, for Lakatos gives no unequivo-cal guidance on how irrecoverable degeneration is to be detected. Since future development cannot be known we can never be sure that some dramatic new reformulation might not occur to revitalize an apparently degenerate programme and transform it into a major success. This, in fact, is identified by many critics as the crucial weakness of the Lakatos account. The decision for or against a research programme appears, as with Kuhn's paradigms, to require an essentially intuitive leap by the individual scientist, even though Lakatos explicitly dissociates himself from Kuhn's 'socio-psychological' framework, describing his own as 'normative'. In any event one critic who makes a good deal of this problem, while simultaneously asserting that Lakatos's theory is the 'most advanced and sophisticated methodology in existence today' is Paul Feyerabend, with whose radical, not to say entertaining and irreverent, views on the philosophy of science, we end this chapter.

Paul Feyerabend and the Principle 'Anything Goes'

The main thread linking all the accounts of science that we have considered so far has been the emphasis on rationality and 'progress'. Despite important differences in just how they interpret scientific advance, all philosophers have stressed the logical processes by which old theories are replaced by new. Indeed, in the accounts of the inductivists and falsificationists, attention is paid almost exclusively to the justification of theories, while their methods of discovery or 'creation' are seen as a matter of interest only to historians, sociologists and psychologists.

In his book *Against Method: Outline of an Anarchistic Theory of Knowledge* (1975), Paul Feyerabend paints an 'irrationalist' picture of science, denies that there is, or ever has been, *an* objective scientific method and claims that if any progress is discernible in science it is the result of scientists having broken every conceivable rule of rationality. Against the background of his more conventional peers, Feyerabend inevitably appears to be extreme. But if he himself uses rhetorical and propagandist methods (with which he characterizes successful scientists) in order to erode our confidence in 'objectivity', it has to be acknowledged that his onslaught is therapeutic if it forces us to look again at the arid solemnity of much of the debate which so sharply divides the philosophers' idealized vision of science from the practitioners' experience of what it is like to do.

Feyerabend starts from the very reasonable premise that 'proliferation of theories is beneficial to science, while uniformity impairs its critical power'. The uniformity that he sees is the product of an ideological conspiracy which is enshrined in the institutionalization of science. Thus present-day science functions much like the Church did in former times, the men in white coats being the modern spokesmen for an absolute, incontestable authority. To stray from the conformist standards of this indoctrinated scientific community is to be branded 'unscientific' and this, in practice, will be construed as advocating not just non-science, but nonsense.

To proliferate theories it is essential to follow 'the only principle that does not inhibit progress . . . anything goes'. That this *is* the only methodology for science is revealed by a scrutiny of history. According to Feyerabend, no episode in real science is sufficiently simple to fit any one of the conventional methodologies; thus it has always been normal practice to cope with inconvenient facts by ignoring them, by explaining them away in blatantly *ad hoc* fashion, or even by concealing them behind a smokescreen of rhetoric. As one example Feyerabend discusses Galileo's defence of Copernican astronomy. Thus Galileo showed through his telescope that the moon had mountains and that the height of these could be estimated by the length of their shadows. By so doing he hoped to refute the Aristotelean idea that the heavenly bodies were all perfect crystalline spheres, yet at the time telescopes were of extremely mediocre quality and the wild inaccuracies of Galileo's own drawings could be seen by the naked eye. Galileo offered no theoretical reasons why the telescopic observations should be accepted as superior when viewing the heavens, and Feyerabend asserts that his only reason for preferring them was that they tended to confirm Copernicus. In short, according to Feyerabend's reading, Galileo prevailed over his critics by virtue of his astute propaganda. He did not show his case to be 'better' but he did present it more skilfully.

Feyerabend's liking for Lakatos's account springs from his claim that it is merely anarchism in disguise. In his failure to define the time limit after which a degenerating research programme must be abandoned, Lakatos seems to admit that there *is* no rational means by which a scientist may direct his allegiance to one programme or another. Lakatos's may seem to be the most sophisticated methodology, yet when stripped to its essentials, it is revealed as no methodology at all. However, this is to be welcomed, says Feyerabend, for not only will the many theories thrown up within an anarchic science have to compete with each other in an atmosphere of unhampered free-for-all, but science itself will also be assessed on its merits in open competition with other avenues of knowledge such as mysticism, astrology and magic. Science cannot be proved necessarily the 'best' ideology for a given individual to follow, any more than a given theory in, or methodology of, science can be proved the best. But by allowing genuine choice (which must mean introducing alternative schemes early in the education system) those who are best suited to science will choose to follow it and in this way science itself can only gain.

(There is further discussion of scientific methods and philosophies, and of their sociological contexts, in chapter 9.)

5

The Nature of Science: Physical, Biological and Social Sciences

In the last two chapters the characteristics and methodologies of the various sciences have been treated as if they were, indeed, all parts of one subject. While it is perfectly possible to defend the position that there is a single philosophy of science applicable across the board to disciplines as diverse as physics, geology, genetics, psychology, sociology and perhaps even some aspects of history, it is also legitimate to argue that there are numerous philosophies of science, each appropriate to its own particular subject matter. As in so many other connections, so with the philosophy of science: it is probably best in an elementary context to avoid extreme positions and to view these apparent alternatives merely as differences of emphasis. It will be the purpose of this chapter to consider these differences as they occur in some of the more important areas of contention. In particular we shall be concerned with the question of whether the differences that undoubtedly do exist between the major types of science, the physical, biological and social, are those of kind or those only of degree.

MODERN SCIENCE AND ITS PHILOSOPHY

We have seen that the most widely influential modern philosophy of science is the so-called hypothetico-deductive model derived in recent times from the work of Karl Popper. According to this view there is no logic of discovery in scientific method and no philosophical importance to be attached to the circumstances in which a hypothesis is formulated. The act of creation is seen as inevitably obscure, although obscurity of origin is not held to affect the claim of a scientific statement to be 'true'. The role of conventional philosophers of science is thus seen as analysis of the logic of justification. In this context the criterion of demarcation between science and non-science is understood to be one of the testability (or more strictly in Popperian terms, the falsifiability) of hypotheses. In the idealized model, a hypothesis is testable in a simple and universal form. It is confirmed (though never verified in the deductive sense) on each occasion that its implications conform to new and varied observations, or disconfirmed by a single instance of nonconformity. Although the search for final proof is futile because logically impossible, the 'truth' of a hypothesis tends to be judged in terms of how closely it approximates to the idealized deductive model – that in which the inference follows of necessity from the premises – of how effectively it explains known data, and how successfully it predicts data as yet unknown. By the same token, an inference which is drawn inductively can be only more or less probable, and hence the explanatory force of the hypothesis is usually weaker.

This type of distinction – simplified as it is – between deductive and inductive inference forms a partial basis for the traditionally hierarchical arrangement of the sciences in terms of the generality of their subject matter. According to this view, physics occupies the most exalted position because its subject matter is universal, being the fundamental physical properties of all materials. Chemistry is somewhat less broad in scope, because while all materials have chemical properties it is usually thought possible to reduce these to the still more basic physical properties. There is a continuum in chemistry from those branches concerned only with non-living systems to those which concern the chemistry of life processes. Thus the biological sciences are less broad again, being occupied only with those bodies which are alive. They also merge gradually into the social or behavioural sciences through the agency of psychology, until in a science such as sociology the scope is significantly limited by its concern with only a small fraction of the world, namely human societies.

Although the reduction of chemistry to physics raises few hackles, there is a continuing, and as yet unresolved, controversy over whether it is permissible to reduce the phenomena of sociology to those of psychology, of psychology to biology, and finally of biology to physico-chemical phenomena (see below). However, the tendency to do so is probably commoner now than at any time in the past. For many centuries the orthodox explanation of natural phenomena was the teleological (i.e. purposeful) one which described present events in terms of future and 'higher' goals. Thus the fall of the apple was the result of the mutual attraction between the earth and the apple's own earthly element; the latter must strive to attain its natural place. Since the seventeenth century, present events have increasingly been explained as the result of the causal influence of past events and this has encouraged the reductionist attitude in science, which is the very opposite to that prevailing in medieval times. This situation raises enormous and wide-ranging philosophical problems, not the least of which is that while we now say that the cause of the apple's falling is universal gravitation, we are able to understand little more about the real nature of this force than could Newton himself. (Indeed, in Einstein's relativity theory gravitation is not regarded as a force at all.)

For many scientists and philosophers it seems that there is a lack of clarity and certitude at the very heart of science. This is a problem of some familiarity to philosophers, but one that is almost universally ignored by practising scientists. The problem is that the fundamental entities in our experience of the world appear incapable of exact analysis or definition, such that at this level there is inevitably confusion and disagreement over the most basic of our conceptual data. So long as this is the case, it will be difficult, and perhaps impossible, to find unique versions of such primary concepts as gravity, acceleration or mass, versions, that is, which command the assent of all. Even the idea of cause itself – one that is deeply entrenched in intellectual history, in the human experience of dealing with the world, and in the mechanistic account of the universe – is now an endlessly subtle and ambiguous focus of philosophical debate.

THE FOUNDATIONS OF MODERN PHYSICS

These problems have been especially acute in those areas of physics concerned with the nature of the material world. In the quest for an understanding of the mysterious phenomena revealed during the twentieth century, physicists have felt increasingly obliged to extend their interests beyond the boundaries of physics, into fields which have traditionally been the domain of philosophy. At the same time philosophers have been obliged to consider the revolutionary discoveries of physics as an important part of the raw data of philosophy. So far these trends have produced no hint of a universal synthesis acceptable to all, but they have at least ensured that the traditional questions of the nature of the physical world and of the meaning and status of life have had to be considered in the light of new evidence.

From the Renaissance to the end of the nineteenth century events in the world were seen increasingly as the predictable consequences of antecedent events, the latter being derivable from clear and simple foundations. This confidence was a reflection of the simplistic and of course hugely successful picture of the world as consisting of solid chunks of a homogeneous stuff which by appropriate organization could be built into machines. Its basic entities were not too far removed from the data of everyday experience. Thus matter was constituted from atoms rather like billiard balls, mass resembled the familiar concept of weight, and the idea of force was visualized by such analogies as the tension developed in muscles.

By the early twentieth century this picture was crumbling from its very foundations. The increasingly tenacious pursuit by the physicists of the fundamental properties of matter revealed entities and concepts quite unlike those of the familiar world. Some of these, such as the seemingly endless series of elementary 'particles' or the so-called waves of probability, seemed unreal, while others, such as the relativity of time and space, seemed bizarre and unnatural. Matter itself was indistinguishable from energy, the electron was not a *thing* charged with negative electricity, but was to be conceived rather as a negative charge charging, as it were, itself.

Far from the nineteenth century's confident belief that the triumphant march of science was somehow to be the final expression of the human spirit, we have now reached a new condition of doubt and uncertainty, a condition in which the only thing that appears to be certain is that our understanding of the universe, and of our place in it, is necessarily limited. In particular the theories of modern physics, coupled with the growth of relativism and the recession of 'objectivity' in philosophy of science, have left the impression that the world so successfully discovered by science is not the world as it really is. Science can, of course, tell us a great deal about the world, but it cannot, seemingly, give us the whole truth.

One of the most arresting discoveries, and that with a particularly disconcerting impact on our overall conception of what the world is 'really' like, was the 'principle of uncertainty' put forward by Werner Heisenberg in 1929. This principle expresses the limitations of the concepts of classical physics in a precise mathematical form. In the minutest world of elementary particles it seems that events do not always follow the strict causal sequence of our normal experience, or of the laws of physics as applied in the macroscopic world. The behaviour of the particles is of course by no means chaotic, yet an apparently inbuilt degree of indeterminacy at the most basic level of function of the world is an immensely radical finding. It means, for example, that it is impossible to measure accurately both the velocity and the position of a fundamental particle, or to say whether or not particles have retained their original characters after colliding. That is, there is an uncertainty both of destiny and of identity. It is extraordinary that in physics, the archetypal 'hard' or exact science, we should find that at this basic level the results of events cannot, after all, be predicted according to the deductive model, but can be specified only in the probabilistic terms usually associated with less precise disciplines. Furthermore, it seems that in mathematics itself, for thousands of years in the Western tradition the very model of intelligibility and rationalism, it is impossible, also in the nature of things, ever to achieve finality of understanding. Thus the 'incompleteness theorem' (1931) of Kurt Gödel showed that no set of logical relations can be established that does not also imply the existence of still other relations with which the set itself cannot cope. Even a pure deductive system is therefore inherently incomplete and uncompletable.

The philosophical importance of these findings is reflected in the immense literature they have generated. While it cannot yet be said that there is unanimity of interpretation ('non-subjective' readings of modern physics have been advanced – notably by Paul Forman – and, of course, also decried), there is no doubt that the impact has been profound for our perception of the world and of the nature of scientific truth. Is the problem merely one of methodology? Will further advances in science enable us eventually to regain a more deterministic scheme of causation? Or must there really be a limit to man's scientific knowledge of the world? If, as many philosophers and physicists believe, the latter may well be the case, these most fundamental discoveries may show us only the blurring of the traditional Western dichotomy between the objective and subjective, the public and private. If in studying nature, science inevitably affects the nature of what it studies, it must be that science after all can give us only the appearance of the world and not its reality. It is as if our plight were no better than that of the prisoners in Plato's Allegory of the Cave. We sit, as it were, chained in one position so that we see only the shadows of 'real' events in the world outside our cave. We can study these shadows with ever greater exactitude, but of the reality behind the shadows we have no direct knowledge.

Heisenberg himself had this to say:

> . . . we can no longer consider 'in themselves' those building stones of matter which we originally held to be the last objective reality. This is so because they defy all forms of objective location in space and time, and since basically it is always our knowledge of these particles alone which we can make the object of science. Thus the aim of research is no longer an understanding of atoms and their movements 'in themselves. . . . From the very start we are involved in the argument between nature and man in which science plays only a part, so that the common division of the world into subject and object, inner world and outer world, body and soul, is no longer adequate and leads us into difficulties. Thus even in science *the object of research is no longer nature in itself, but man's investigation of nature.* Here again, man confronts himself alone.

We have come far since the seventeenth century when Galileo could say with such confidence: 'The conclusions of natural science are true and necessary, and the judgement of man has nothing to do with them.' We have come, in fact, so far as to have arrived at the point where the interests of science coincide with perhaps the greatest question in the whole of philosophy: the problem of knowledge itself. What can we *really* know? Ultimately, it appears to many modern thinkers, science cannot study nature 'in itself', but only human investigations of nature which, incidentally, implies a good deal about the philosophical preferences and prejudices that individuals must surely bring to this study (see chapter 6). We cannot examine the behaviour of the elementary particles but only our knowledge of this behaviour, and it is in this sense that Heisenberg concluded that 'man confronts himself alone'. We might conclude this brief comment on an enormous topic with a fine (though, of course, highly controversial) passage quoted by Heisenberg from his fellow-physicist Arthur Eddington (1882–1944).

> We have found that where science has progressed the farthest, the mind has but regained from nature that which the mind has put into nature. We have found a strange footprint on the shores of the unknown. We have devised profound theories, one after another, to account for its origins. At last, we have succeeded in reconstructing the creature that made the footprint. And Lo! it is our own.

THE IDEAS OF BIOLOGY

This new image of man and of his perception of reality, has been derived exclusively from the physical sciences and from mathematics. The picture would be incomplete without some consideration of the contribution made by the life sciences, most particularly because the idea that chance and uncertainty exist as basic elements in the workings of nature is more familiar to biologists than to physicists. It is implied in the theory of organic evolution – the great organizing principle of biology – which was earlier in origin than quantum mechanics and relativity, and in many ways has implications which are just as far-reaching.

The biologist is thoroughly accustomed to processes that are developmental. He encounters development both in what he calls ontogeny and in phylogeny, that is, in the apparent 'progress' of the individual through its long and complex embryological growth, as well as in the changes through geological time that he observes in populations of organisms. Thus it is that he examines the exquisite intricacies in the development of the chick embryo, and observes in the fossil record the 'horse' as it descends through tens of millions of years from a dog-size creature to modern *Equus*. At both levels the change can be seen as the progressive realization of potentialities. And it is a selective process because it yields, out of the original undifferentiated cells of the embryo, cells specialized to perform particular functions only (muscular, nervous and so on) in the adult; and from a population of 'generalized' ape-like ancestors, 'specific' forms such as the chimpanzees, gorillas and men of the present age.

Again at both levels, it seems that information contained in potential form in a simple system ultimately achieves realization in a more complex system. However, perhaps the most fundamental distinction between the processes of embryology and those of evolution, is that in the former normal development probably does not involve chance factors, whereas in the latter chance mutations are almost certainly essential in the long term. There is no doubt that, given sufficient generations, significant changes do occur in viable populations as a result merely of the recombination of genetic potentialities during interbreeding. 'Natural' selection by the environment of any favourable characteristics which may result is thought to ensure that organisms possessing them leave more offspring. But it would be absurd to believe that the animal kingdom in all its variety has evolved simply by the shifting and sorting of the hereditary potentialities present in the original population of single-celled organisms. The 'spontaneous' appearance of new potentialities by mutation gives to evolution that extra dimension over embryological development which accounts for the appearance of organisms not present, even in potential form, when the process of change began (see also below).

It also seems to imply that in biology these chance factors are the creative basis of the value judgements we make about such concepts as 'improvement' or 'progress'. Although such a contention quickly takes us into philosophical areas of intense controversy, not to say of shifting fashions, it would be difficult to argue that man is not in some critical way an improvement on his ape-like forebears, or that the chicken is not an advance on the egg. What is more, these judgements appear to arise from the biological system itself rather than from any external notion derived from ethics. As such they perhaps imply the acnowledgement of a more positive interpretation of the significance of the chance happening in living systems than in the sub-atomic world of the physicists. In the first case chance seems to be the source of 'improved' and more complex organization, in the second of disturbance to an otherwise precisely determined system (see chapter 7).

The Organized Whole

Despite contrasting interpretations, these developments in physics and biology have led in their separate ways to another movement in modern science of immense interest and importance. This is the study of organization *per se*. It is perhaps odd that the investigation of the fundamental properties of matter should have helped to stimulate this, but it did so in two ways. One was through the reappraisal of our understanding of the atomic nucleus which is now seen, not as a body composed of elementary particles held together by attractive forces, the particles and forces being quite separate entities, but rather as if the particles and forces were different and complementary aspects of the same thing, appearing separately only when the essential organization of the intact nucleus has collapsed. The other contribution of fundamental physics was the impact of quantum mechanics on the traditional separation between what is being observed and what is making the observation. If it is impossible to make observations at this level without disturbing the objects of that activity, then what we are in fact studying is the organization constituted from the two components. We have returned, in fact, to the famous dictum of the Greek philosopher, Parmenides (*c*.540 BC): 'The All is One'.

At the level of biology the concept of 'one-ness' or the organization of the whole is familiar enough in studies such as embryology. But it is applied much more widely than this involving, as it does, the crucial idea of the inter-relatedness of component parts, such as is seen in the normal functioning of the animal body. The stability that results from inter-relatedness in physiological systems is referred to as homeostasis. A simple example would be the control mechanism that governs body temperature in man, which in recent years has been described in terms derived by analogy from engineering systems. Here we encounter expressions such as 'set point', 'thermostat' and 'negative-feedback', by which the components of the body are compared with mechanical or electronic components of the engineer's control system. The essential feature is that the desired stability is achieved by these various parts working as an organized whole. In exercise, when the muscles produce more heat than normal and raise the body temperature, this information is fed back to the brain via the blood stream so that the temperature of the thermostat registers higher than the set point. The resultant 'load error' ensures that appropriate information is now transmitted to the 'effector mechanisms' which organize the relaxation of peripheral blood vessels and sweating. When these mechanisms have dissipated the excess heat and the temperature has returned to normal, they are shut off and the whole system resumes its resting state.

The point illustrated by this example is that organization at any level, from that of sub-atomic particles to human societies, involves more than mere relationships between component parts of the system. It implies the maintenance of a 'preferred' stable-state against influences from within or without which tend to upset the stability. The organization thus operates self-protectively as if it were a complex 'whole', exhibiting properties that are qualitatively different from those of its individual components.

Teleology

The homeostatic system maintaining constant body temperature also provides us with a model around which we can briefly discuss the problem of teleology, one which has been of central interest in biology for many centuries. As we have seen, teleological modes of expression characteristically refer to the ends to which things allegedly strive rather than, as in causal modes, to the antecedent events that brought them about. Although the original Aristotelian conception has long since been abandoned, biological systems often appear to be teleological, and biologists habitually speak in teleological terms. The question to be asked therefore is whether the phenomena of biology are qualitatively distinct from those of the physical sciences to the extent that they need different modes of expression. A simple example will make the problem clear. From his knowledge of the gas laws, a physicist might say that raising the temperature of a gas causes its volume to increase. Similarly, a physiologist might say that raising the temperature of a dog causes its respiration to increase. However, although the physiologist might conclude that the function of the increase in respiration (panting) is to oppose the rise in temperature (i.e. to keep the body temperature constant), the physicist would be most unlikely to say that the maintenance of thermal equilibrium was also the 'function' of the increase in the volume of the gas.

The question of under what conditions the use of teleological modes is justifiable has no easy answers and need not concern us in any detail. Many philosophers of science prefer to present the statement 'B is a function of A' as 'A is a cause of B' and certainly no one would now suggest (as Aristotle did) that teleological language would be appropriate for describing the workings of the solar system. Nevertheless, in the case of living organisms, teleological terminology is likely to be with us for the foreseeable future. And between these two extremes there are inevitable grey areas.

One aspect of this difficult subject is illustrated by our example of the temperature control system, because most of the essential features of the physiological system can be duplicated in a simple physical analogue, namely the system governing the temperature of a room. In this system are two of the most basic characteristics traditionally used for distinguishing teleological and non-teleological organization, those of 'preferred-state' and 'negative feedback'. Yet the system governing room temperature has been 'designed' and is entirely non-living. Its preferred-state is the temperature at which the thermostat has been set and it is the task of the system as a whole to maintain this temperature in the room. This is done (figure 4) by having available 'effector mechanisms' (analogous to shivering and sweating) for heating or cooling the room air according to the direction in which its temperature fluctuates from the preferred state. The feedback in this system would be the room temperature prevailing at any moment, which the thermostat compares with the preferred temperature; it is said to be negative feedback because a deviation from the preferred state initiates mechanisms which tend to oppose the deviation.

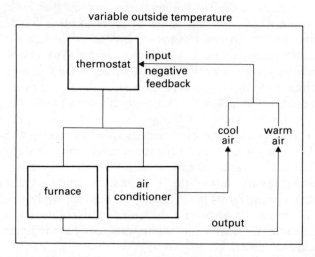

Figure 4   *The control of room temperature: a diagram illustrating the principles of preferred state and negative feedback*

In simple terms it seems that our ability to draw fairly close analogies between physiological and physical control systems takes a good deal of the mystery out of the concept of teleological organization and satisfies for most people the claim that 'goal-seeking' systems need not necessarily be living.

Reductionism and Emergence

Our discussion of organization and teleology now leads us to consider the wider problems of reductionism. The life processes once again take a central position here because they raise the question of the nature of their differences from non-living processes. What do we mean by saying that living organisms differ from inanimate objects in having different types or levels of organization? Can the life processes be 'reduced' to physico-chemical processes without losing the essential features that make them living? Is there any support nowadays for the idea of the operation of some vitalistic principle which makes living systems unique?

To begin with it must be said in relation to all these questions that the central debate, whether in the traditional guise of 'vitalist versus mechanist' or in the modern form of 'organization versus reduction' has not been, and probably cannot be, resolved exclusively in terms of empirical evidence about the world. Before we look at the bare essentials, all that need be said about the voluminous literature on the subject is that whereas the original positions taken up by the protagonists were clear and simple, the current situation is one of enormous sophistication. The result of this appears to be that while the passion of the debate may be undiminished, the two sides are now significantly closer than at any time in the past, a situation to which the intricate subtleties that have become necessary features of the arguments bear ample witness.

The problem of reductionism in biology arose out of the historical fact that the sciences of physics and chemistry provided the life sciences, in their infancy, with a sound basis from which they could develop. But how far could this basis be relied upon? René Descartes, the so-called father of modern philosophy, is generally held to have orginated the particularly rigid distinction between 'mind' and 'matter' that has characterized or bedevilled Western thought ever since. Because man's body was of the material world, it was to be explained in the reductionist terms of material entitites and mechanistic principles. The subjective world of the mind was, however, something altogether different. Descartes' own doctrine of mind was in fact somewhat confused but, roughly speaking, mind and body were said to be fully independent, not interacting as such but proceeding, as it were, in parallel. Although Descartes' view was subject to a number of interpretations, the early vitalists came to believe that the properties of living systems could be explained only by the operation of some non-material agent over and above the properties of the physical body.

In these terms the debate was clear enough, and it raged furiously throughout the nineteenth century. The beginnings of a reconciliation, or of a clouding of the opposing issues, began in the early years of the present century with the origins of modern physics. The vitalistic claim that life processes differ in essence from the mechanisms of the material world, implies that we actually do understand what materials and mechanisms are. This was all very well in the deterministic world of billiard-ball atoms, but in the age of relativity and quantum mechanics the idea of a mechanistic account of the world became for many scientists a good deal less plausible. The result of this was that it seemed less clear upon what basis the vitalists and the mechanists could debate.

One of the dominant thinkers of this period, and one of the founders of our modern interest in relational and organizational properties, was the mathematician and philosopher A. N. Whitehead (1861–1947). Whitehead attempted to resolve the vitalism-mechanism problem by developing the so-called doctrine of emergence, first advanced by C. Lloyd Morgan (1852–1936). He began his analysis, not from our understanding of the ultimate constituents of the physical world, but from the evidence about the world derived from direct observation of it. We pass, that is, not from knowledge of the basic entities of the world to knowledge of how they account for the properties of living organisms, but rather from observable phenomena to our intellectual constructs about the basis of matter. Our knowledge of these basic entities is necessarily imperfect and mysterious because we can know only those properties that are manifest in the phenomena in which they participate, and which we can observe. Thus we should not hope, as it were, to recognize in isolated atoms of oxygen and hydrogen the peculiar properties that permit them in appropriate combinations to make water; but when the properties of water do 'emerge' from this combination, we have indeed observed something new about the atoms, something that we can learn only when they are *organized* in this particular way.

So it is with life. When we recognize that atoms of oxygen, carbon, nitrogen and so on, when organized in a particular way, manifest the properties associated with living systems, we do not need to fall back on the idea of adding some vital force, any more than we would do so to account for the appearance of magnetism in an iron bar, or the ability of a radio tuner to bring us the piano sonatas of Beethoven from the Wigmore Hall. What we do need to understand is that these impressive properties spring from certain organizations of material systems; in the absence of these organizations the properties are no longer manifest. In these terms, life itself loses much of its mystery (at least in the philosophical sense) and it is no longer such a problem to account for its appearance in the evolution of organisms from inanimate materials, or of its disappearance on the death of the body.

What, finally, can be made of the mind-brain problem itself, a puzzle which has exercised the greatest thinkers throughout history? There have always been those who believe that all material phenomena, including those of living bodies, may in principle be understood in terms of physics and chemistry, but who nevertheless have insisted that subjective phenomena are of a different kind. There are still some eminent advocates of this view, philosophers and scientists among them, who point out that despite the very considerable knowledge now available on measurable events in the brain, events which are unfailingly associated with mental phenomena, we are no nearer to closing the gulf between an event such as an electrical discharge and our subjective awareness of an abstract thought or an emotion in the mind. To say that they are *identical* seems to beg the question, and it is in any case known from the work of the psychoanalysts and others that a great deal of 'mental' activity proceeds in the brain without our being aware of it. Subjective (i.e. conscious) experience is not an inevitable correlate of nervous processes, even of some processes which we should regard as thought if they reached the level of consciousness. Rather, it seems, consciousness may be something unique.

Although we cannot here be concerned with the several major philosophical problems suggested by these findings, the phenomenon of *self*-consciousness is of special importance in connection with what was said above about organization. If self-consciousness exists in man, and if we believe in evolution, from what did this phenomenon evolve? Is it possible to account for the problem of 'mind' in much the same way as we accounted for magnetism and life, that is, as an emergent property of certain types of organization in material systems? The idea that something of the same qualitative kind as self-consciousness may occur in sub-human forms of life is not difficult to comprehend, but the logical implication of this way of thinking would appear to be that an analogous phenomenon might occur also in inanimate organizations and ultimately in atoms themselves. This will seem unintelligible to many, yet it has not seemed so to some influential thinkers, among them Whitehead himself, the biologist J. B. S. Haldane (1892–1964) and the palaeontologist and Jesuit priest Pierre Teilhard de Chardin (1881–1955) (see also chapter 8).

Another, and sharply contrasting, view of the mind-brain problem that we should mention briefly is the reductionist theory known as behaviourism. This school of psychology tries to describe the phenomena of 'the person' in terms of the phenomena of the behaviour of the body. The rationale behind this is that while specifically mental events are not denied as such, they are to be excluded from the realm of scientific enquiry on the ground that they are not available as 'public knowledge' for all to scrutinize. Attention is to be directed to the overt forms of behaviour, which are seen either as the public expression of psychological phenomena or in some way as actually equivalent to them. In any event, it is asserted, the terms conventionally used in psychology to describe mental states and processes are to be understood as ways of describing the associated forms of behaviour in given situations. Thus persons who are said to be clever, stupid, kind, dishonest and so on, may be observed to act or to have a tendency to act in ways that we should regard as characteristic of these respective adjectives. Other experiences 'in the mind' can, it is argued, be dealt with in much the same way, such that the idea of the 'mind' itself becomes redundant for the purposes of psychology. According to this view there is no more point in seeking to understand how mental events influence the behaviour of the body than there is in asking how your car's reliability affects its mechanical perfomance.

The behaviourist position is open to a number of important objections quite apart from the incomprehension or repugnance that it engenders in some individuals. For example, a little thought will show that we are forced to use traditional psychological terms as well as behaviourist terms in order to determine if a person's behaviour is, let us say, stupid or dishonest; we need also to know, among other things, something of his attitude of mind, his beliefs, desires and so on.

Finally it is perhaps worth adding that neither the enormous interest in organizational 'wholeness' nor in reductionist relations has led to any overall unification of science itself. Science cannot easily be viewed as 'one', simply because its various disciplines may be said to deal with the 'emergent' properties of disparate organizations. To be sure, there have been heroic attempts at reductionism between the major divisions of science, as well as within them, but the successes have been few when the effort has even been worth making. A famous recent example has been the attempt to reduce traditional Mendelian genetics to molecular genetics, which is a specialized type of biochemistry of nucleic acids. This treatment received a good deal of support in the limited areas where it proved possible, yet it left untouched far more of the original subject than it was able to redefine, including evolution itself which appeared incapable of any such redefinition. Of course it is always possible to argue that at some time in the future, when our overall understanding is far greater than it is today, a truly unified science will emerge, but for the foreseeable future it appears that the study of organized systems has stimulated an appreciation of the different techniques and intellectual processes appropriate to the different levels of complexity. The system of the particle physicist is likely to remain orders of magnitude simpler than that even of the molecular biologist, while the sociologist's subject matter appears so complex as often to defy our comprehension.

THE SOCIAL SCIENCES

The problems raised in the biological sciences by teleology and reductionism lead us to the admittedly 'purposive' phenomena studied by the social sciences. There is much impassioned debate over the attempt to reduce the propositions of sociology to those of psychology, but the arguments too often appear motivated predominantly by the threat to professional identity. No doubt there are biologists who would wish to reduce even these threat reactions themselves to simple biological imperatives, yet this sort of controversy is trivial by comparison with the genuinely fundamental questions considered above and with those which we must consider below.

The reductionist implications of the new inter-discipline called 'sociobiology' is conceivably an exception. Its advocates predict that the impact of biology on the social sciences and humanities will in the long run be as revolutionary as that of chemistry and physics on biology. The suggestion that social science will one day be considered a branch, as it were, of biology, flatly contradicts the prevailing wisdom which postulates that human social life is almost entirely the product of cultural influences, exhibiting little or no evidence of genetic determinism. But this is a wide ranging and seemingly intractible debate and for our present purposes it is sufficient to note that the conception of biology as the key to understanding human behaviour, is just the latest contribution to the nature-nurture argument which has raged, unresolved, for many centuries.

We may now consider a matter which, among some natural scientists, is considered only slightly less controversial than the above, but which against the background of our earlier comments concerning the nature of scientific knowledge should pose few serious problems. This is the question of whether it is, after all, possible to tackle social and human phenomena by means of the 'scientific method'. (It is important here to dissociate clearly between 'methods' and 'techniques' in the context of science. By 'the scientific method' is meant an identifiable version of the somewhat elusive procedure described, for example, in chapters 1 and 3. 'Scientific techniques', on the other hand, are particular means for implementating the scientific method – means such as surveys, measurements, controlled experiments, and so on – which vary widely in accordance with particular sciences.)

Is it not the case, then, that the sheer complexity of human life dooms to failure any attempt to be genuinely scientific?

Complexity of subject matter is undeniably a problem, for when there are many interrelated phenomena it is difficult to establish any clear causal sequence. In its extreme form this view has it that human life is so rich and subtle that every social event is unique; any such belief, sincerely held, would be incompatible with scientific investigation (see the beginning of chapter 1). Yet can this view really be sustained, or at least can it be shown that complexity is uniquely the problem of the social sciences? There is no doubt that the biological sciences share many similar difficulties, more especially those branches such as ethology and ecology which study the behaviour of organisms in natural and semi-natural conditions. Even in the physical sciences, there is also formidable complexity, and the response to this has always been first to accumulate knowledge of the simpler phenomena, in the hope that they will eventually throw light on the more complex.

Some variety of this simplifying process seems to be applicable across the enitre spectrum of the sciences, and the ideal approach is undoubtedly to achieve simplification by setting up artificial, controlled experiments. However, experiments are not always possible and here it is interesting to recall that the dominance of astronomy in the seventeenth century was certainly not the result of its being able to manipulate the celestial bodies experimentally. Similarly, many modern and highly reputed natural sciences, such as geology or evolutionary biology, have little scope for experimentation, yet they have not failed to establish extensive bodies of knowledge, with well-founded general laws and theories. Hence those areas of human social enquiry in which opportunities for controlled experiments are rare, cannot be disqualified from the ranks of science on this account alone. In any event there are some areas, notably social psychology, where experiments indistinguishable in design from those of the natural sciences are routinely performed, while economists make extensive use of idealized models which may be analysed mathematically in much the same way as in physics or physiology. Finally, the 'field investigations', in many social sciences, do not differ in any significant way from those in, say, botany or entomology.

Another alleged difficulty for social science concerns generalization. Whereas a typical law in physics or chemistry is usually thought to have universal application in the sense of being independent of space and time, human affairs tend to be influenced significantly by cultural or historical factors. Behavioural norms are thus unlikely to be valid between different societies or within a given society at different times. This tends to mean that social theories, although they often have considerable explanatory power, seldom provide much basis for accurate predictions of the future. This deficiency might appear a crippling disadvantage if comparison were now made with positional astronomy, but a little reflection quickly shows that astronomy is the exception rather than the rule, even among the physical sciences. Accurate prediction of future states is possible, even by means of the famous physical laws, only within certain artificially idealized conditions (for instance, in a 'perfect' vacuum, or given 'perfect' rigidity) which are analogous to the idealized market forces applied to the equations of economists. With the less exact sciences, such as meteorology, prediction is notoriously hazardous, while with living systems (not to say subatomic physics) we are seldom dealing with anything better than probabilities.

A wide-ranging and genuine problem for social science is that concerned with the so-called relativity of laws about human societies. Can the study of human phenomena ever be fully objective and value-free? For example, in psychology and social medicine, there is often the issue of whether we can investigate or experiment upon people and yet retain our respect for them as human individuals. Equally, there are certain sorts of phenomena which are difficult to study because the necessary measurements may be dangerous or unacceptable, or the data may have to be treated with confidentiality. Indeed, it is the very closeness of such sciences to urgent personal matters of daily life which can reveal the values and prejudices of the scientist himself in terms of his decision as to what to study in the first place. If his decision is based largely on reasons external to the science itself, (as, let us say, in the case of a wish to 'prove' a contentious point) rather than on widely acknowledged intra-scientific ones, the investigation itself is less likely to satisfy normal scientific standards of impartiality.

Where questions of moral judgement are concerned (as for example in a study of race prejudice or sexual behaviour) the situation is particularly sensitive and obviously very different from that in which a chemist or physiologist selects a problem which, for him, excites a special intellectual curiosity. In the social context there is far graver risk that personal values may bias the very collection of evidence by their influence on what is considered to qualify as factual information, and then intrude into the discussion and interpretation of the findings made. It is difficult for the social scientist to achieve genuine detachment on delicate human issues, the more so if, as is by no means uncommon, his sense of intuition as a layman seems to provide a deeper, more direct insight than his formalized methodology. Unfortunately, our common-sense judgements are frequently misleading, and it is just because these reflect attitudes to issues far wider than those strictly relevant to a given area of social research, that the problem is so intractable. Of course in the natural sciences, too, there are many areas of controversy, and always some that generate passionate discord, yet the resultant tension is often a creative one from which genuine solutions may spring. This is simply because natural scientific debates typically concern the interpretation of exclusively intra-scientific data, whereas those in the social sciences invariably draw upon more general viewpoints.

This problem can in some contexts seem so insoluble that it is occasionally recommended that social scientists abandon the attempt at value-free analysis and rather specify their own attitudes as fully and honestly as possible, so that others can evaluate their importance independently. This is in many ways a laudable objective, but since it would generally be impossible to know if it had been fulfilled, a completely value-free social science may be unattainable. A controversial response to this realization has always been to argue that such a social science is in any case undesirable; social phenomena must always have an essentially 'subjective' or 'value-impregnated' aspect, because of the nature of purposive human action. According to this view, the attempt to exclude subjective interpretations inevitably eliminates also every genuine social fact. Consequently the so-called 'non-objective' techniques of enquiry should be developed, to include in particular the willingness and ability of the social scientist to project himself empathetically into the phenomena he is studying. This alone will lead him to hypotheses with real explanatory power.

Yet another problem facing the social scientist is that knowledge of social phenomena is itself a social variable. This means simply that the human subjects of experimentation or investigation are likely to behave abnormally if they are aware of what is happening to them. For example, people may answer survey questions in terms of what they think is expected of them, rather than of what they actually believe or do, or they may even give false information out of some wish to obscure the results. Furthermore, people tend to believe that they 'know' about social trends and attitudes, such that even after an exhaustive analysis of some social phenomenon by scientists, they continue to doubt or reject conclusions contrary to their own. This situation would be regarded as highly eccentric in connection with the physical sciences, although it occurs on occasions in biology, and particularly of course in those applied branches such as medicine which are closer to immediate human concern.

Human reactions to knowledge of social phenomena affect predictions as well as explanations. This has been illustrated in recent times in connection with the attempt to forecast the outcome of elections. If the prediction influences people's voting behaviour at all, it may do so either by encouraging them to act in accordance with it (the 'bandwagon effect') or contrary to it (the 'under-dog effect'). If the social scientist attempts to allow for these effects he can expect success only when they are equal. If he makes only conditional predictions which do not allow for the reactions to the forecasts, the initial conditions will not be fulfilled. That is, once he attempts to influence the outcome of a prediction, he cannot then check its reliability.

Finally, then, if we ask the question: Is social *science* really possible? we must formulate our answer so as fully to acknowledge the undoubted problems, but at the same time to recognize the objectives of the whole enterprise. Human nature does present some formidable difficulties to any who would attempt to explain, let alone predict. Yet explanation and prediction are among the legitimate objectives of social science just as they are of natural science, and it is beyond dispute that there is already a substantial body of literature which fulfils, in general terms, all the normal criteria of scientific knowledge. Although there still remain too many examples of pompous, vague and ambiguous social scientific prose, in which the understanding achieved and transmitted may be no more profound or reliable than that of the traditional common sense judgements of folk wisdom, it must be said that such lack of clarity is by no means the monopoly of social scientists. It is probably true to say that areas of vagueness (in the data rather than merely in the prose style) will always present special difficulties for social scientists, but this having been acknowledged, there seems every reason to expect that the problem areas will continue to yield to responsible analysis and that as a result we can look forward to a continued expansion of real understanding.

Part II

Interactions of
Science and Society

6

Social Studies of Science and Technology

The purpose of the present chapter is to examine science as an institution of the modern world. This means that we must turn scientific method on to science itself in much the same way as a biologist, for example, might describe and analyse an organism, identifying its salient characteristics, the environmental conditions that promote its growth, and the behavioural patterns by which its future development might be predicted. In the early sections of the chapter we are therefore concerned with the elements of science that can be 'measured'; the features that are said to give to the scientific enterprise its internal cohesion; and the processes by which scientific knowledge grows. (An account of the Marxian approach to science is given in chapter 8.) The later parts of the chapter describe a dynamic new school of sociology of science which first emerged from an investigation of the links with technology, and also consider matters of enormous practical importance in the modern world: the industrial and military applications of science and their consequent impact in economic terms.

QUANTITATIVE AND QUALITATIVE ASPECTS OF SCIENCE

Any attempt at a 'science of science' will tend to emphasize aspects which can be objectively analysed and measured. The sorts of parameters taken as measures of science might include the numbers of people practising it, the amount of money spent on it, or the quantity of 'knowledge' it produces. Any of these could be expected to give an idea of how science has grown in the past and, perhaps by extrapolation, of how it might be expected to grow in the future. They are therefore likely to be thought important by those concerned with science policy.

Exponential Growth
The most influential work of this type has been that by Derek de Solla

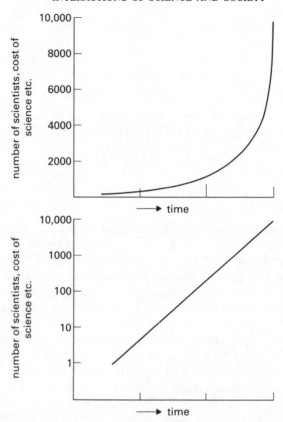

Figure 5 *Exponential growth, plotted linearly and semi-logarithmically*

Price, started in the 1950s and brought together in his widely-quoted book, *Little Science, Big Science* (1963). The first so-called 'law' to emerge from Price's studies was that the growth of science has been exponential (figure 5):

> Our starting point will be the empirical statistical evidence drawn from many numerical indicators of the various fields and aspects of science. All of these show with impressive consistency and regularity that if any sufficiently large segment of science is measured in any reasonable way, the normal mode of growth is exponential. That is to say, science grows at compound interest, multiplying by some fixed amount in equal periods of time.

It should be stressed that growth of this kind is in no sense peculiar to science; however it *is* what might be expected because it is a common characteristic of a host of phenomena, including the inorganic (for example, the rate of a chemical reaction with rising temperature), the biological (number of bacteria, fruit flies or people in a freely reproducing population) and the social (gross domestic products of nations).

What is of special importance about the exponential growth of science is the short time required for a doubling of size when this is measured in any reasonable way. For example, whereas the doubling time for world population has been about 50 years, and that for a nation's gross national product typically about 20, scientific publications or memberships of scientific institutions have doubled in number in each period of about 15 years. Price examined scientific journals in particular detail. Since the first appearance of the *Philosophical Transactions of the Royal Society* in 1665, the number of publications has fitted the exponential curve quite closely (figure 6). With a doubling period of 15 years, the

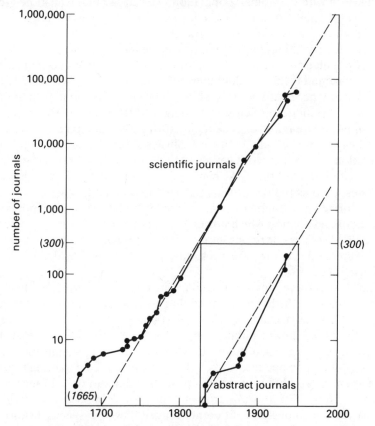

Figure 6   *The exponential growth of scientific journals since the seventeenth century*

number of publications has grown by a factor of 100 every century; thus for every single paper published in 1665, there were of the order of 100 in 1765, 10,000 in 1865 and 1,000,000 in 1965 (but see the next section).

During each interval of 15 years there will appear almost as many new scientists again as there were in all the preceding period of growth. Furthermore, since the 'typical' scientist might have a working life equivalent to three of these doubling periods (45 years), seven scientists will be alive $(1+2+4)$ at the end of his career for every eight there have ever been (that is 87.5 per cent). This phenomenon is called by Price the 'coefficient of immediacy' of science; even if the doubling period were to vary from 15 to, say, 10 years, or 20, the coefficient of immediacy would vary only from 96 per cent to 81 per cent respectively. It is therefore this short doubling time which gives to science its voracious growth; by comparison with the human population, we can say that whilst the great majority of scientists who have ever lived are alive now, most of the people who have ever lived are now dead.

Generally speaking, the growth of science has altogether outstripped that of population; every doubling of the latter has been accompanied by approximately three doublings of the former. Two qualifying remarks are necessary at this point, however. The first is that the immediacy of modern science is nothing new. It is merely the result of three-hundred years of exponential growth. As Price puts it, 'Science has always been modern: it has always been exploding into the population. . . . Scientists have always felt themselves to be awash in a sea of scientific literature.' As compared to the history of other human activities, such as politics or war, this might be taken to suggest that the history of science ought to concentrate on the work of the previous generation, a situation which however is very far from the case. Some of the reasons for the enormous importance that is attached to the study of the seventeenth century are presumably clear to the reader who has proceeded thus far.

The second qualifying remark concerns the generality of the 'law' of exponential growth. Even assuming that we can satisfactorily overcome such problems as deciding *who* is to be counted as a scientist, or *what* is to qualify as a scientific publication, it has to be acknowledged that the statistical approach soon breaks down when anything significantly less than a 'global' perspective is taken. It is known, for example, that there have been widely different rates of growth within different branches of geology and physics. It is also true that the growth of science as a whole has varied enormously in different countries, where special historical factors have been influential. If the output of scientific papers is taken as a standard of comparison, it happens that the Soviet Union contributed less than 1 per cent of the world's total in 1910, but 18 per cent by 1960. During the same period the British Commonwealth countries contributed a constant 13–15 per cent, while Germany's share fell from 40 to 10 per cent. These discontinuities introduce a cautionary note, and perhaps a salutary one for those concerned with matters of decision in science policy-making. Price's persuasive curves mask a very significant degree of averaging-out; within individual sciences and individual countries, there is nothing inevitable about the course of 'progress', nothing, that is, which cannot be influenced by policy judgements. Consequently there are grave dangers in any attempt to treat the exponential curves as genuine 'laws' by which the future may be determined.

With this important proviso, it nevertheless remains true that if the *overall* exponential character of scientific growth has some of the properties of a law, it is better for our understanding of the institution of science if it is not just an empirical law, but one with some modicum of theoretical support. Just as with a biological growth process, the explanation of scientific growth is to be found in terms of two factors, its internal 'vitality' and its external environment. On the one hand the vitality of science derives from the fact that the answers to questions which are yielded by scientific enquiry themselves pose new questions; science thus has the capacity to generate investigations of ever-increasing scope and the potential for endless acceleration. On the other hand the external environment of science is, of course, the society within which it operates and with which it interacts. An unfavourable social environment can inhibit, and even strangle, science before its own momentum produces the positive feed-back which then influences events to its own advantage. The necessary positive influence can be seen for the seventeenth century in the ever-expanding cultural role of science as a substitute for religion. Thus the more the world came to be understood in scientific terms, the more need there was for scientific education and the more was human progress seen as a function of scientific advancement. With the rise of industrialization science became a vital component in commerce and in war, changing the very nature of society and making itself indispensable to modern life. In this way it was science which created the conditions necessary for its own continued growth.

The Logistic Curve
No more than a glance is needed at figure 5 to realize that exponential growth cannot continue for ever. If, as Price shows, science has grown by five orders of magnitude in 150 years, it clearly cannot go on doing so. If it were to grow by another two orders of magnitude we should (say

by the middle of the next century) have 'two scientists for every man, woman, child and dog in the population'. What happens in practice is that the exponential growth reaches some 'natural' limit (in the case of science, shortage of resources and manpower) and the growth curve, instead of increasing inexorably in slope, begins to flatten out with the impending saturation. This S-shaped form is known as the logistic curve, of which the exponential curve is only a part (figure 7). Once again it is not peculiar to science, but accurately describes the growth of biological populations through time, or that of an individual plant with age. It is by comparison with the development of a beanstalk that Price suggests that the long period of exponential growth of science is akin to its exuberant juvenile phase. At the present time, however, science is beginning to enter a more stable phase of adult life. The transition from pure exponential growth to exponential growth with saturation (see figure 7) is one of prolonged crisis and, according to Price's data we are now, in

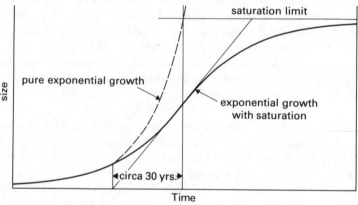

Figure 7   *The logistic curve*

the 1980s, approaching the mid-point of this crisis which extends for about a generation on either side of the point of inflection of the curve. His description of the characteristics of the crisis period (written at the beginning of the 1960s) sounds a now familiar note:

> There will be rapidly increasing concern over those problems of man-power, literature and expenditure, that demand solutions by reorganiz-ation. Further, such changes as are successful will lead to a fresh escalation of rapid adaptation and growth. Changes not efficient or radical enough to cause such an offshoot will lead to a hunting, producing violent fluctuations that will perhaps smooth out at last.

In view of the 300-year conformity of science as a whole to the exponential growth curve, it does seem that the falling away from this pattern (saturation), which began to be clearly perceptible in the 1960s, marks something radically new in modern science. In Price's terms, it marks the transition from little science – that performed by individuals or small groups of co-workers – to Big Science – that of the giant research organizations, working with huge teams of scientists to long-term planned programmes.

The experience of working in Big Science is quite different to that of the long-established tradition of the lone pioneer. Because Big Science is in its first generation, it is inevitable that most of its personnel were trained, and given expectations, in the old-style atmosphere. This can be one of the chief sources of tension within Big Science, for it raises conflicts of loyalty between widely different sub-cultures. The old-style scientist does not take readily to the new demands for bureaucratic research administration and management, being inclined to believe that the 'conveyor-belt' atmosphere of team research, in which each individual performs no more than his own limited skill, is incompatible with intellectual freedom, originality and fulfilment. Moreover, he feels that in the hands of his successors – those who have not experienced something of the traditional craft-training of the young scientist – science itself may lose its unique inner logic, its imaginative and critical independence being swamped by the ever-closer identification with industrial 'missions'.

These grave problems of our present period of crisis must, says Price, nevertheless be seen as 'the beginning of new and exciting tactics for science, operating with quite new ground rules'. The logistic curve suggests new opportunities for future development such that present-day Big Science may prove to be a

> brief interlude between the traditional centuries of little science and the impending period following transition. If we expect to discourse in scientific style about science, and to plan accordingly, we shall have to call this approaching period New Science, or Stable Saturation; if we have no such hopes, we must call it senility.

The Scientific Elite

In the two previous sections we were concerned exclusively with quantitative aspects of science-as-a-whole. But science is ultimately the work of individual scientists and, as individuals, there seems no reason why they should not vary in much the same way as individual members of any other group of human beings. Are there, then, any sound ways by which the *quality* of individual scientists may be judged?

According to Price, 'the gap between the Nobel Prize winner and a plain person is rather larger than that between an Olympic gold medallist and an ordinary mortal'. Within science itself, one admittedly crude but by no means entirely subjective method of assessment, might be in terms of the number of papers published. On this basis, 'productivity' or 'achievement' turns out to vary by a factor of 1000:1, for the most prolific scientist appears to have been the British mathematician, Arthur Cayley (1821–95), with 995 papers (22 per year for 45 years!), while to warrant the title 'scientist' at all we might reasonably decide that it is necessary to have published at least one. Of course, this means of assessment could be said to beg the very question it claims to resolve because it disregards the *quality* of the papers produced yet, given the problems inherent in any such assessment, a variation in productivity of this order is thought to apply equally to other forms of creative activity. While the quantity of published material is not an infallible indicator of the quality of the creator, creators of widely accepted quality are typically prolific.

Investigation of the number of published papers in science shows, in fact, that there is a very uneven distribution across the population of scientists as a whole. Contrary to one's expectation of a 'normal' statistical distribution in which most scientists produce a number of papers close to the mean, it turns out that a small 'elite' fraction (about 6 per cent) of all scientists produce half of all the published science, and the top 1 per cent produce almost a quarter. The elite few – Price estimates them at about the square root of the total number of practising scientists – constitute the creative heart of what is often called the Invisible College of those people who really matter for the advance of a given discipline.

The case for the existence of such Invisible Colleges is relevant in connection with the somewhat traumatic transition from little science to Big Science discussed in the previous section. We saw that a characteristic trend within Big Science is the appearance of team research. When research is performed by teams rather than individuals, work that is published appears under the names of numerous authors instead of just one or two as was almost universal before the present century. Although the trend towards multiple authorship began well before the era of Big Science, it has in recent years become the norm in areas such as high energy physics. Here it is common to find even two or three *dozen* authors listed as responsible for one paper, a phenomenon which not only renders the crude paper-counts of an earlier era hopelessly inapplicable to any assessment of the merits of individuals, but also seriously undermines the more subtle, if subjective, bases of such judgements. Group collaboration in research is, in fact, tending to exaggerate even further the discrepancy between the minority of high achievers in the Invisible Colleges and the mass of scientists who are minimally productive. This is because the former group can increase their output still further by leading research teams, while the latter degenerate into mere 'fractional' authors, who produce no more than 1/nth part of one paper.

Big Science, team research, and multiple authorship together, thus seem to be undermining the traditional role of the scientific paper which has been built up over three centuries. Even the latest tool of 'quality' assessment, the Science Citation Index (which was not, of course, introduced with any sinister purpose in mind), may be of limited value in this respect. This index lists, under a given paper, all those later papers which cited it in their own bibliographies. In principle, it thus allows some idea of the importance of a given paper to be gained by the frequency with which it is referred to in subsequent publications. Needless to say, there are innumerable problems of interpretation in this superficially simple procedure, because there are many quite different reasons for citing a particular paper in connection with one's own work, by no means all of which have to do with its scientific importance. There have been many cases, for instance, where a paper breaking fresh ground has soon been superseded (and no longer cited) because of a rash of new publications for which it was responsible, whereas a paper reporting no more than a refinement of a technique is often quoted, almost in ritual fashion, for many years.

While the publication of scientific papers is likely to remain a useful guide to the growth of science as a whole, it may therefore be of declining value as an indicator of the merits of particular works. This is another symptom of the state of crisis of modern science. The growth of the international, highly mobile, Invisible Colleges within each major research area has made the traditional scientific paper 'an art that is dead or dying', for communication is increasingly by direct contact, from person to person. The role of the research paper has already become secondary in the most dynamic areas because the delay in publication (which can be as long as two years) ensures that when it does appear, the paper can do no more than confirm what the members of the elite circle already know. This tendency works to increase the power of the Invisible Colleges both within science and between science and the social and political forces of its environment. So excessively 'undemocratic' a situation, however natural or inevitable it may seem, obviously raises a host of important questions for science policy and education.

STRUCTURAL AND FUNCTIONAL ASPECTS OF PURE SCIENCE

In addition to the quantitative approaches to the structure of scientific literature, two other important themes have arisen in the sociology of science. The first to appear historically was that concerned to examine the 'external' influences on science or, perhaps more accurately, the reciprocal relations between science and other social institutions. This approach was closely associated with the names of Karl Marx and Max Weber, and it sprang from an earlier and broader intellectual tradition which we now refer to as the sociology of knowledge, a discipline that explored the influence of social forces upon the origins and development of ideas in general. Marx's view was that the system of ideas identified with a particular society was most powerfully influenced by the society's economic base, that is, by the prevailing economic conditions of life. Thus it was an individual's position in the class structure, for example, which largely determined his ideas and beliefs (see chapter 8). Weber, on the other hand, while not discounting the importance of purely material factors, argued that ideas themselves also have an important function in society. It was for this reason that he examined the general world outlook and value-system of Protestantism in respect of its impact on the development of capitalist economics in Europe.

Weber's approach was taken up and extended by other sociologists in a number of studies of the relations between science and religious values or political systems, most notably perhaps in R. K. Merton's so-called 'Puritan thesis', as presented in his *Science, Technology and Society in Seventeenth Century England* (1938). Although Merton did not explicitly identify the causal influences underlying the rise of modern science, he clearly implied that the dominant set of *values* prevailing at the time – in particular those which encouraged practical experiment – as distinct from any material factors, were chiefly responsible. Merton's thesis found such fertile ground in the Western academic world that it was for almost a generation accepted with little criticism, stimulating several other studies of similar kind.

Despite this, however, the tradition inherited from the broader sociology of knowledge was thought by many scholars to be too vague for satisfactory analysis, so that the main influence of Merton's early work proved to be in promoting not the 'realistic', but elusive, study of science as an integral part of the wider society, but rather the internal analysis of an 'idealized', but accessible, science, defined as a more or less independent social subsystem. In this approach the methodological and cognitive aspects of science are accorded much less prominence than the behavioural norms of the people who carry it out. Because it emphasizes the internal workings of science as a social institution – for example the social characteristics of the organization of science and the commitments of scientists to a clearly understood theoretical framework – the approach is often referred to as 'functionalist'. That is, it concerns the forces which function to maintain the *status quo* of the subsystem of science. Functionalism represents the second of the two major themes mentioned above, and as examples we shall briefly examine the later work of Merton himself, and that of T. S. Kuhn, together with the extension of their ideas in the hands of W. O. Hagstrom and M. J. Mulkay.

The Ethos of Science

Merton's own work within this second main theme constitutes what some still regard as the only mature tradition in Western academic sociology of science, even though in recent years it has been under increasing attack. It presents a highly idealized conception of science in terms of four norms, or institutional imperatives, which are said to govern scientific activity. These imperatives are binding on scientists not only because they are technically efficient as the best available means of attaining science's overriding goal of advancing knowledge, but also because they are believed to be 'right and good'. In other words, they are ultimately moral prescriptions.

Merton's first institutional imperative is that of *universalism*. The purpose of this is to guarantee that new knowledge is evaluated solely in terms of objective, impersonal criteria, and that such things as career advancement are determined solely by talent. Thus, he says, 'the Haber process [for the preparation of ammonia] cannot be invalidated by a Nuremburg decree nor can an Anglophobe repeal the law of gravitation.' Universalism is subjected to the most acute strain at time of international conflict, for it is incompatible with values such as 'ethnocentralism' or nationalism. On the other hand, it is fostered by democracy, since this values accomplishment above mere status.

'*Communism*' is the second norm, intended in the sense that the 'goods' of science – its body of public knowledge – are subject, not to private, but to common or public ownership. Science accumulates knowledge by means of its extended, co-operative enterprise, its findings being assigned to the whole community. As a result the only property rights of the individual scientist are those of recognition and esteem which accrue roughly in proportion to the importance of his work. It is for this reason that disputes over priority of discovery, a recurring phenomenon throughout the history of science, are seen as conforming entirely to the institutional imperatives. The right to be acknowledged for good work is absolute. Conversely, secrecy by the scientist is forbidden. It is the individual's duty to communicate his findings among his colleagues, because the most basic institutional goal of science is the extension of knowledge. Thus the individual must act always for the common good, for he himself has benefited from the communal efforts of his predecessors, as Newton himself acknowledged in his famous remark, 'If I have seen farther it is by standing on the shoulders of giants.' And despite the competitive nature of science, the 'common good' extends beyond the confines of any particular laboratory to the scientific community at large. This is the reason for the conflict between the idea of 'discovery' in science – something that may bring distinction to an individual or his group, yet which is available for all to use – and that of 'invention' in technology, where the rush for patents so clearly proclaims exclusive rights of use, the very antithesis of communism.

Merton's third imperative element of the scientific ethos is that of *disinterestedness,* taken to mean a deep, but detached interest in the workings of the world, an interest that is, simply for its own sake. The scientist is expected to make his discoveries available for public scrutiny and there is, accordingly, little scope for fraud or irresponsibility. In this important sense the behaviour of scientists is subject to close institutional control. A failure by the individual to conform will be condemned by the wider community, thus enforcing the translation of disinterestedness into practice, no matter what motives may operate unseen. While science shares this norm with other professions, there is less scope for the scientist to exploit the ignorance or credulity of the layman than there is in the case, let us say, of medicine or law. As a result, science and scientists have traditionally enjoyed a particularly high status for reputability and ethical behaviour (but see chapters 8 and 9).

The final institutional imperative is what Merton calls *organized scepticism.* Clearly, this interrelates with the others and, indeed, is a crucial part of scientific methodology itself. It is this element which most often brings science into conflict with other social institutions, such as the Church or the State, for science 'does not preserve the cleavage between the sacred and the profane, between that which requires uncritical respect and that which can be objectively analysed'.

Social Control Within Science

In Merton's work the question of deviance from the norms laid down by the scientific ethos was little examined because it was considered to be so unusual. The question of motivation was largely ignored. In a direct extension of Merton's theoretical framework, W. O. Hagstrom has however stressed these two aspects in interpreting his empirical study of academic scientists. In recognizing deviant behaviour, he came to emphasize those institutional conditions which tend to minimize it. For Hagstrom, the outstanding motivation of pure scientists is the desire for recognition by their fellows and in order to achieve this they offer 'gifts' or 'contributions' to the scientific community in the shape of publishable information. There is thus said to be an exchange of information for recognition at the heart of the institution of pure science.

While the nature of the information offered will be determined by the ethos of science, the recognition sought may come in several ways. In the first instance it comes through the publication of papers in the learned journals; 'status as a scientist can be achieved *only* by such gift-giving.' The appearance of papers indicates the attainment of the standards of the community, because the best scientific journals publish only a carefully selected proportion of the communications they are offered. Further, and more tangible, recognition may follow in the form of election to specialist societies, opportunities for attendance at conferences both at home and abroad and, ultimately, for the select and lucky few, perhaps membership of one of the great general scientific academics, or even a Nobel Prize.

If the exchange of information for recognition is to work as intended within the 'pure' atmosphere of academic research, it must not be corrupted by influences from outside. Such influences might include certain other, competing, exchange-systems identified by sociologists, such as those involving political or economic pressures. For example, if a scientific journal operated, let us say, by a commercial publisher instead of a learned society (there are many such journals), were to lower its refereeing standards merely as a means of increasing its 'share' of the discipline in question and with a view to increasing its sales, it might attract papers for the wrong reason, scientists quickly seeing this as a way of advancing their claims for recognition. Any such corruption of the scientific exchange system would prostitute the ethos of science and be seen in terms of the involvement of unwelcome factors from outside of science itself. According to the orthodox view, the ethos of pure science can be sustained only so long as it operates on internal factors alone, that is, so long as it is fully autonomous.

Hagstrom sees it as the function of the exchange system to 'internalize' the values of the scientific ethos. This is achieved by preserving them from the corrupting influence of external factors, specifically by rewarding conformity and punishing deviance. A good deal of conformity is always assured by the protracted 'socialization' process through which scientists must pass during their training, but it is only the exchange system which can eliminate any deviance that might still occur. Like Merton, Hagstrom emphasizes the frequency of priority disputes as evidence in support of his theory, for it is here that competition for recognition is at its fiercest. Without acknowledgement of priority the system of incentives would be threatened, and it is the strong desire for recognition that induces scientists to undertake the chore of writing-up their results, in conformity to the essential norm of communism.

Not only does the desire for recognition induce the scientist to communicate his results; it also influences his selection of problems and methods. He will tend to select problems the solution of which will result in greater recognition, and he will tend to select methods that will make his work acceptable to his colleagues.

This, says Hagstrom, is how the 'differentiation of disciplines' sets in, the creation of new specialties being the most effective means of avoiding intolerable competition for limited rewards. Such dispersion can lead to isolation, both geographical and social, and Hagstrom draws the analogy with competitive behaviour in the organic world, specialization among scientists being roughly equivalent to speciation in nature. The specialty is the environment to which the specialists become best fitted and which largely determines their behaviour. In order to understand scientific behaviour it is accordingly essential to have some appreciation of particular specialties. Hagstrom's move here, away from an exclusive concern with the Mertonian norms, reflects the influence of modern historians of science who are concerned to identify those characteristics of scientific knowledge itself which induce change, rather than the social forces acting on this process. They are, in a word, interested in how science grows.

Social Processes of Innovation

We have already seen that T. S. Kuhn's book *The Structure of Scientific Revolutions* has been acclaimed as a major intellectual landmark with the widest application (chapter 4). This is confirmed by the fact that its influence on the sociology of science has, if anything, been even greater than that on its philosophy. Kuhn's work has provided a new and subtler insight into the internal processes of science which account both for its stability and for its growth. Having discussed his main ideas in chapter 4, we shall here be content merely to relate them to other sociological studies.

According to the Mertonian tradition, scientific knowledge will accumulate by virtue of conformity to the institutional norms. For Kuhn, however,

> successful research demands a deep commitment to the *status quo,* [though] innovation remains at the heart of the enterprise. Scientists are *trained* to operate as puzzle-solvers from established rules, but they are also *taught* to regard themselves as explorers and inventors who know no rules except those dictated by nature itself. The result is an acquired tension, partly within the individual and partly within the community, between professional skills on the one hand and professional ideology on the other.

Within 'normal science' the Mertonian imperatives may perhaps operate, but the primary guidance for a given field of research would be that
field's prevailing paradigm; recognition for the scientist would come
from its refinement and articulation. Normal science is, in Kuhn's
words, 'the attempt to force nature into the conceptual boxes supplied
by professional education.' It is only within these conceptual boxes that
detailed investigation of problems suggested by the paradigm is
possible, and it is they which in due course generate the anomalies that
lead to re-structuring of the accepted model, or even to its revolutionary
overthrow. This indeed is how Hagstrom's differentiation of disciplines
comes about, for the creation of specialties encourages the growth of
knowledge by concentrating the specialist's attention on problems which
are viewed by his community as both important and soluble. Their
solution brings him the recognition that he seeks.

The British sociologist M. J. Mulkay has used the 'Velikovsky affair'
to illustrate his claim that in Kuhnian revolutionary science – the crisis
period after the accumulation of anomalies which leads to the abandon-
ment of one paradigm and the acceptance of another – the Mertonian
norms do not operate at all. The cosmological work of Immanuel
Velikovsky, first published in his extraordinary book, *Worlds in
Collision* (1950), purported flagrantly to challenge many of the fundamental assumptions of biology, geology and astronomy. We do not need
to examine the details of the case in order to understand that the
assumptions challenged were the prevailing paradigms of their
respective disciplines and, in Mulkay's analysis, the violent, even
hysterical reaction of scientists that was wrought by Velikovsky's book
indicates that their behaviour was governed, not by a simple allegiance
to the moral prescriptions of the scientific ethos, but rather by a powerful sense of psychological commitment to theoretical and methodological norms. Had the Mertonian imperatives been effective, Velikovsky's
claims would have been subjected to detailed, disinterested, and critical
examination. Instead there were vicious attacks on Velikovsky as an
individual; judgements on his book apparently accompanied by refusals
to read it; and attempts by the scientific 'establishment' to prevent
unrestricted access to Velikovsky's own data. These 'unscientific'
reactions quite clearly broke the norms of universalism, organized
scepticism and communism. For Mulkay, they re-affirm Kuhn's thesis of
the rigidity, rather than the flexibility of the scientist in his attachment to
paradigms and indicate that it is this rigidity rather than the Mertonian
imperatives which guarantees the growth of knowledge.

Mulkay does, however, extend Kuhn's ideas to produce a scheme
which, rather better than Kuhn's own, renders compatible this intellectual inertia within science with the rapidity of actual scientific advance.

Mulkay notes that growth in science occurs as a result of the spread of a paradigm's influence from the area in which it was formulated into other areas. This is achieved by virtue of the research community's being composed of a host of overlapping problem networks. These represent areas of smaller intellectual scope than the conventional disciplinary boundaries which have been institutionalized within the university framework, their members being specialists whose professional concern is with a quite strictly confined set of problems. The importance of such networks lies in the fact that new research findings must first attain the status of genuine scientific knowledge by a process of validation by the members of the relevant specialty. Such validation bestows a degree of recognition on the scientist(s) responsible and further improves the cohesion of the research network in terms of the consensus attained. Those who make findings of widely recognized importance emerge as the elites of the specialty; their success attracts new entrants to the field and this in turn promotes exponential growth.

When this growth becomes saturated, important new findings get increasingly rare. The professional rewards of recognition then become inadequate and recruitment to the specialty declines. Moreover, those individuals established within it begin to look elsewhere in order to avoid stagnation, migrating into new 'areas of ignorance' where they can apply their former experience to problems which have yet to achieve scientific consensus. Because of this, and because in a new field there are as yet no committed specialists, there is less inertia in newly developing areas and less resistance to innovation. At any rate, it is in this fashion, according to Mulkay, that old research networks degenerate and dissolve while new ones take root and grew.

The actual process by which new ideas are generated is, says Mulkay, closely tied up with the 'cross-fertilization of ideas'. As evidence of this he points to the contrasting structural and functional organization of French and German science in the nineteenth century, the one rigid and heavily centralized, the other loose and competitive. Cross-fertilization is promoted by a fluid system which encourages a degree of role-duality on the part of individuals, as in the case of 'practitioner-scientists' or of 'researcher-teachers'. Such flexibility helps to loosen the 'mental sets' which, as Kuhn contended, are a consequence of the traditional pigeon-holing system of education in the sciences. Innovation then results from the creative blending of otherwise divergent frames of reference.

Although working ostensibly within the 'internalist' theme that developed from earlier functionalism, Mulkay is clearly aware that in recent years, with the fall off in the exponential growth curve for science as a whole, much has changed. Criticism of science – particularly that from 'inside' – is more widespread than ever before (see chapter 8).

Even allowing for the short-fall in financial support, recruitment of trained personnel has become inadequate. Above all, constraints on the scientific research community have come increasingly from non-scientific directions, notably from a lay public which demands more accountability in terms of political and economic goals. If these trends continue – and there is every expectation that they will – the convenient separation between an idealized realm of pure science which hitherto has exerted so exclusive a hold on the interests of orthodox sociologists, and that of a science applied to industrial and military objectives, must inevitably become less distinct, as well as less tenable. It is therefore now time for us to examine some of these matters for ourselves. (For reasons that will become obvious, the emergence of a thoroughgoing non-Mertonian sociology of science is described in chapter 9.)

SCIENCE, TECHNOLOGY AND INDUSTRY

Understanding the social organization of pure science is obviously important, but it must not be at the expense of the study of what is more representative of science as a whole. In the twentieth century, typical science has become the 'applied' science that is performed in the government or industrial laboratory. The schemes of Merton, Hagstrom, Kuhn, Mulkay, and most of the other sociologists, can be applied – if at all – to no more than a small minority of all scientists, perhaps to fewer than 10 per cent. And even within the framework of pure science, when this becomes Big and teamwork the order of the day, it is in any case impossible to fit it to such models as the information-recognition exchange system.

An important reason for this distinction between pure science on the one hand and applied science and technology on the other, is the differences they exhibit in the sociology of publication. Whereas the pure scientist is strongly motivated to publish his findings because this is the only way in which he can hope to establish new knowledge and, with it, his own reputation, the technologist most often lacks access to anything equivalent to the scientific paper. This is not to say that there is no technical literature, which is far from the case, but rather that the literature which exists serves a different purpose. According to Price, this purpose is

> a newspaper-like current affairs function, for boasting and heroics, and probably, above all, as a suitable burden to carry the principal contents of advertisements which together with catalogues of products are the main repositories of the state of the art for each technology.

There are no publicly-accessible papers in the scientific sense because the technologist is not communicating knowledge openly but is often concealing knowledge in order to gain commercial advantage over his competitors. Unlike the scientist, he has no Invisible College through which the latest knowledge is transmitted long in advance of published work. Consequently he is far more eager to scan whatever literature is available in order, as it were, to glean useful information by reading 'between the lines'. Price summarizes all this in the aphorism: 'the scientist wants to write but not read, the technologist wants to read but not write' and defines technology 'as that research where the main product is not a paper, but a machine, a drug, a product or a process of some sort'.

Unfortunately, our tendency to speak in this way of science and technology, implying that they are quite separate activities, has led to a good deal of confusion. To some extent this is the result of outdated and unclear thinking. (We will provide clear definitions for the contemporary situation in the next section.) For example, many writers in seeking to determine the historical relations between science and technology, have presupposed just such a clear distinction between a concern with theoretical knowledge on the one hand and with the practical arts and crafts on the other. Within this convenient perspective the sorts of questions that can be tackled include: Which of the two comes first, science or technology? Are they causally related? Can any consistent pattern be uncovered? Needless to say, this approach has yielded few simple answers. Amid a host of controversies, the one agreed conclusion to emerge is that there is no simple model which governs the relations in every case. In the pages of history examples may be found where a utilitarian technique appears to have arisen as a linear consequence of fundamental scientific research; equally, examples are available in which practical technological advances led to new scientific understanding; where theory and practice developed hand-in-hand; or apparently in total independence.

For our purposes, there is little point in jumping into the controversy itself. We are fortunate that the traditional histories of science and technology, which often emphasized the heroic discoveries and inventions of a few Great Men, in isolation from the social forces of their day, have in recent times given way to a more sophisticated analysis which places such creative activities firmly in the context of cultural and philosophical dimensions. One point has emerged as quite clear and beyond dispute. This is that during the later nineteenth century – whatever pertained before that time – there was a very marked convergence between science and technology, manifest chiefly in the increasingly conscious and increasingly widespread application of scientific knowledge to technical practice. This continued apace, and by the early twentieth century, technology and industry had become tightly linked together in a complex web of mutual *inter*dependence.

In the modern world, any vestige of confusion may best be dispelled by seeing the divisions between science and technology in terms of the social environments in which they are carried out. It is not simply that science values knowledge while technology values actions; rather it is that the communities which support science and technology tend to value 'knowing' and 'doing' respectively. By the same token, the audience for science tends to be constituted of other research scientists, whereas the main audience for technology is composed of non-researchers who seek results of practical utility. The history of technology in the nineteenth century shows us that its convergence with science was not restricted merely to the absorption of the latter's factual knowledge but, perhaps more importantly, of its experimental methods and institutional framework as well. We see this, for example, in agriculture, where the great agricultural societies began to employ chemists, botanists and entomologists to work on problems of vital practical importance, as well as in engineering which likewise established professional societies, published learned journals, and developed research laboratories along the lines of those in physics. The technologies thus themselves became scientific and, in addition, gave rise to the specifically technological sciences whose only difference from their 'purer' relations is that they pursue the sorts of knowledge most closely associated with practice. Again in the case of agriculture, it is clear that the technology itself has grown more scientific with every decade in this century; it is served by a host of technological sciences, from soil physics and crop genetics to livestock production and microbiology, while the whole enterprise is conducted under the umbrella of agricultural economics which tries to maintain a thoroughly realistic contact with the wider world. Similar examples apply with medicine and engineering.

The dynamics of the modern system ensure that relevant information can flow in both directions between the two wings of what has become the research-and-development continuum. This is not of course to say that the channels of communication are perfect or, indeed, that their mode of operation is well known. Much undoubtedly remains to be discovered by future social studies of science and technology, both within the ideal environment of the large, multidisciplinary laboratory, and beyond. But at least we can already say with confidence that the traditional model of science as the originator of knowledge, and technology as the means of its application, no longer helps us to understand the current position; technology is not simply *based* on science.

To improve our understanding of how the two relate, we must now examine further the role of science in modern technological industries.

Pure Science, Applied Science and Technology

Although we have taken pains in the above section to show that any rigid distinction between science and technology can seldom be justified, in referring to the research continuum we have avoided suggesting that clear differences of emphasis do not exist at the extremes. Obviously a technology such as agricultural engineering requires very considerable skill of hand and eye, but not a particularly deep theoretical background, while a science such as mathematical physics may be exclusively theoretical and demand no more apparatus than pencil and paper. In the more contentious areas it is futile to engage in impassioned debate, not least because our notions of science and technology are in any case not fixed, but constantly changing. Thus there can be no definitive view as to whether the landing of men on the moon, or the development of nuclear reactors, were triumphs (or disasters) of science or of engineering. They clearly involve the complex interaction of both, so that an insistence upon a separation between the scientific and the technological components of the whole process will be little more than a declaration of personal taste.

It is fortunate that a good basis for making such distinctions as are useful is available in a publication of the Organization of Economic Co-operation and Development (OECD) entitled: *The Measurement of Scientific and Technical Activities* (1970). This distinguishes between basic research, applied research and experimental development. 'Basic research is original investigation undertaken in order to gain new scientific knowledge or understanding. It is not primarily directed towards any specific practical aim or application.' 'Pure' basic research is performed at the will of the individual scientist, while 'oriented' basic research is steered in a general way by the individual's employing organization towards some field of particular interest.

'Applied research is also original investigation undertaken in order to gain new scientific or technical knowledge. It is, however, directed primarily towards a specific practical aim or objective. Applied research develops ideas into operational forms.'

Technological or 'experimental development is the use of scientific knowledge in order to produce new or substantially improved materials, devices, products, processes, systems or services.'

These, then, are the separate components of the total activity which in the twentieth century we have come to know as Research and Development. This activity is defined by the OECD as 'creative work undertaken on a systematic basis to increase the stock of scientific and technical knowledge and to use this stock of knowledge to devise new applications.' The whole enterprise is summarized diagrammatically in figure 8.

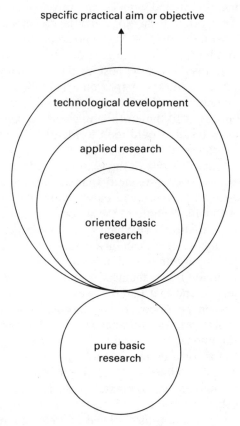

Figure 8   *Diagrammatic presentation of the concepts of basic and applied research and technological development*

An example given in the OECD publication to illustrate the three major categories in the research and development process is taken from the medical world. Thus the investigation of the sequence of amino-acids in an antibody molecule would be regarded as *basic* research. The use of this information to distinguish between the antibodies of various diseases would be *applied* research. The final phase of *technological development* would then be concerned with finding a technique for synthesizing the antibodies for particular diseases and with testing their effectiveness, perhaps in clinical trials.

While these separate categories are still routinely distinguished in theory, it cannot be stressed too strongly that in practice the whole process is continuous, involving at all stages the contribution of individuals of similar background and expertise. Thus a modern industry may incorporate individuals who were trained as basic scientists but who perform work in technology and development, and others who were trained as technologists yet who function as scientists; the intellectual procedures followed and the results obtained, are frequently indistinguishable. Just how we choose to classify such workers may tell us more about the criteria employed than about the individuals' activities. For example, a classification based on community member-ship may encourage us to think of them as technologists; one based on the publication of results would tend to identify the scientists. At any rate, this arbitrariness of division is characteristic of more and more modern research and it is one of the reasons why technology is so much better regarded as 'science-related' than as science-based.

Even the so-called pure basic research, which is typically performed in the disinterested atmosphere of the supposed academic ivory towers, often turns out (as in the above example) to have practical applications. Because the history of science can turn up numerous such examples, it is not surprising that arguments about the 'unforeseen uses' of pure research are common when financial support is needed. No one, for example, would pretend that Faraday was not pursuing knowledge 'for its own sake' when he discovered electromagnetic induction. When asked what use his work was, Faraday is said to have replied, 'What use is a baby?' With hind-sight we tend to see his experiments as one of the essential first steps towards the technologies of 'wireless' communica-tion and electrical engineering. Much the same could be said of Einstein's celebrated equation, $e=mc^2$, a rarified product of theoretical physics if ever there was one. Yet it 'led' to the development of nuclear weapons, and in this case Einstein lived to see his 'pure' work 'abused' in practice. The example of the discovery of the mould, *Penicillium* (1928), is rather different. Although made by Alexander Fleming in very much the spirit of detached, theoretical science, he was working in the bacteriological laboratory of a hospital, and was thus not cut off from practice. Fleming's work was 'oriented basic research'; he could see the mould's potential in medicine, but lacked the facilities to exploit it. In the event his work was developed into the anti-biotic industry very largely by a pharmaceutical company, and in this way pharmaceuticals became a classic example of a science-related industry where there has always been constant interaction between theory and practice.

Similar examples could be repeated at will. But if on the level of methodology we can recognize quite clearly that the categories within the whole research and development enterprise are increasingly 'flags of convenience', on the level of economics the distinction does remain relatively clear. To those providing financial support it is still technology which, potentially, can make money, whilst science always costs money. Much 'pure' science, and not a little 'applied', makes no headway in the limited sense of economic return; but in the main, investment in technology begins only when the best economic indicators point to a profitable outcome. (A notable exception such as the super-sonic aircraft project which produced Concorde, may be regarded as a technical success that nevertheless remains a financial disaster; it merely goes to show that the 'highest technology' is often supported for reasons of national status or security rather than for economic benefit, a subject considered later in the chapter.) Comparatively speaking, very little money is spent nowadays on science for purely 'cultural' reasons, in the way that it is spent, for instance, supporting the arts. The great majority of scientific activity in the modern world is undertaken in the hope of gaining economic or military advantage, and it is the chief remaining function of this chapter to consider these aspects of its social dimensions.

The Emergence of Science in Industry

The roots of today's science-related, high-technology industry may be traced back perhaps to the time of the Renaissance, to the practically-orientated philosophy of Francis Bacon and to the experimental method of Galileo. It was only then, with the opening of the new age of exploration, that mariners could turn to the natural philosophers for guidance in the accurate determination of such essential parameters as latitude and longitude. By 1600, a publication such as William Gilbert's *De Magnete,* which is famous chiefly for its work on the basic properties of magnetism, also included substantial sections on the use of magnets in nautical instruments and on practical difficulties in the mining and working of iron. Bacon's own concern was particularly with the application of experimental science to such 'industrial' arts and crafts as these, and it was in large part on the basis of his philosophy that the Royal Society of London, with its consciously utilitarian attitudes, was founded in 1660.

By the end of the seventeenth century, the pressing practical need to substitute coal for the rapidly diminishing supply of timber, had seen the development of Thomas Savary's 'Miner's Friend', the first practical steam engine to be applied to the problem of drainage. During the next two generations, by the introduction of far-reaching design improvements, some of them based on fundamental thermo-dynamics, this first, leaking, and highly inefficient monster was transformed into the universal workhorse in the nation's burgeoning factories. The advent of steam power was thus a consequence of continuously shifting relations between technology and science, and it played perhaps the major role in what Eric Hobsbawn has called 'the most fundamental transformation of human life in the history of the world recorded in written documents.' It was a social upheaval of the most daunting complexity and we are fortunate in having as our objective no more than the briefest thumb- nail sketch of its founding influence on the modern industrial research- and-development laboratory.

The principal technological innovations of the early Industrial Revolution (from approximately 1760) were wrought by unlettered craftsmen, canal builders, millwrights, ironmasters and blacksmiths, working in the countryside. The truly revolutionary character of their inventions and improvements lay in the fact that for the first time they were designed, not for the privileged few, but for the everyday use of the common people (as well, of course, as to make a profit). In a remarkably short time they led to the substitution for animal and human labour of mechanical devices driven first by water and later by steam. (The emphasis on this trend is in no way intended to obscure the appalling conditions which quickly resulted from the new domination of people by the pace of machines.) The rapid acceleration of industrial output, which was a main characteristic of the period, came largely from the coupling of these simple technical advances to the profound organizational change of bringing numerous small productive units – which for countless generations had been scattered in rural villages – together into large urban factories. This system required a clearly structured division of labour, with greater discipline, regularity and routine; it diffused 'rational attitudes' and 'scientific methods' through- out industry and beyond.

The process of industrialization occurred first in Britain, and the key sector in the beginning was without doubt that of cotton textiles. Raw cotton could by this time be imported cheaply from the new American plantations; there were rapidly expanding markets for manufactured goods, not least those at home; and the re-organization of the time- honoured techniques along new lines quickly proved capable of satisfying the demand. Thus the chief driving force behind the technical changes was economic, not scientific. The production of cotton goods rose by five times between 1766 and 1787, a period which saw the introduction of those justly famous innovations, the spinning jenny, the water-frame and the hybrid mule which dramatically accelerated the spinning of cotton thread, and the flying shuttle and power looms which revolutionized weaving. The machines themselves owed nothing to theoretical science, and in this respect the textile industry was repre- sentative of the early industrial revolution as a whole.

By the second half of the nineteenth century, however, the balance was changing. The production of cloth depends upon other processes in addition to the mechanical ones of spinning and weaving, processes such as bleaching and dyeing. It was these which provided opportunities for the systematic involvement of scientists with specific theoretical knowledge. The case of the Germans synthetic dye industry is particularly instructive. Although the English chemist William Perkin (accidentally) produced the first artificial, fade-resistant dye, 'Mauve', from derivatives of coal tar, it was the Germans who realized the economic potential of the discovery and invaded the field in a big way. By 1900 they were in complete control, supplying some 90 per cent of all synthetic dyes throughout the world.

Superficially this extraordinary success was the result of the Germans' superbly organized, economically-motivated industrial machine. More fundamentally, it reflected their determination to overcome a late start in industrialization by directly applying their now-pre-eminent scientific expertise. Their ability to do this was in turn a consequence of the dynamic, competitive atmosphere generated in the German universities during the nineteenth century (in contrast to general 'Laissez-faire', amateurism and complacency in Britain), and specifically of the high quality and quantity of their teaching and research in organic chemistry under the pioneering leadership of Justus Liebig, himself devoted to the ideal of seeing his chemistry applied.

Another example was the development of the electrical industry. The discoveries of the pioneer scientists, Davy, Faraday and others, took some time to find application, and in an important sense the steam engine–that nineteenth-century symbol of 'progress' – was a party to the practical developments in electricity which eventually did come. The engine was quickly adapted for use on locomotives and, later, for driving electrical generators. The invention of the electric telegraph (1837) was important for signalling on the railways and it soon found use in general communications. The ultimate extension of this was perhaps Thomas Edison's invention of the electronic valve, which paved the way for 'wireless' contact, communication of this type being first achieved across the Atlantic in 1901.

The first commercial applications for the dynamo, a direct offspring of Faraday's experiments, were in electro-plating and electric lights. The latter were used initially in isolated places such as lighthouses, where a generator could replace batteries. By 1879, however, Thomas Edison in America and Joseph Swann in England independently developed the evacuated wire-filament lamp from the early lead given by Davy. This breakthrough created an unprecedented consumer demand, promoted work on the means to distribute electric power, notably that in the

General Electricity Research Laboratory (1906) in the United States, and instigated the boom in electrical industries which within a few decades were satisfying multifarious and hitherto undreamed-of applications.

By the turn of the century the examples of the electrical and chemical industries were no longer exceptional; more and more widely the lesson of the German initiative had been learned. In the future science would be the indispensable partner of technology in industry, and any who failed to make tactical and strategic use of this new situation could expect to pay a heavy price in the increasingly competitive commercial world.

Technological Science

The convergence of science with technology on a large scale began, then, with the appearance of the industrial research-and-development laboratory towards the end of the nineteenth century, a trend started in Germany and followed rapidly in the United States, and then in Britain and other countries. Applied, industrial research was made possible by the emergence of the 'professional' scientist, one who had received a prolonged, specialized training and who for the first time was acknowledged to have a socially important role to play. The research and development laboratory was the effect, not the cause, of closer links between technology and science, although once accepted it dramatically accelerated their interdependence. Industrial symbiosis thereafter became self-perpetuating because it was increasingly hazardous for governments and commercial companies to be left behind in the competitive race for technically new processes and products. These, it was quickly realized, would come only through the innovative activity of organized scientific and technological research.

The results in the twentieth century of this near-union of science and technology are too numerous and too well known to need repeating. It will be sufficient if we briefly consider the examples of the aircraft industry and the development of nuclear power.

A whole generation of piston-engined, propellor-driven aircraft evolved from the Wright brothers' first sensational test flights in North Carolina (1903), but it was the imperative of national defence which finally broke down the conservatism of the aero-engine designers of the late 1930s and encouraged work on the gas turbine. The first British design was projected in 1928 by a young trainee engineer, Frank Whittle, who wrote that 'the turbine is the most efficient prime mover known' and could be 'developed for aircraft, especially if some means of driving a turbine by petrol could be devised'. It is significant that Whittle's basic plan drew extensively on his own theoretical knowledge of thermodynamics, so that he was able to apply fundamental principles of physics to the details of design. Furthermore, his engine required the use of the new heat-resistant alloys which were only then being produced by the relatively new science of metallurgy. But Whittle's ideas could obviously not have been implemented without substantial financial backing and technological expertise. There was a long series of technical problems and near-disasters before the first British jet flight in 1941. By this time the Germans had built a similar engine independently, and it seems that their two-year advantage over the British, which might have been decisive in the progress of the war, was lost through inadequate technological support.

The technology of the aircraft industry, and its relations with the science of aerodynamics, is in many ways a modern parallel of the nineteenth-century story of the steam engine and thermodynamics. The empirical origins of the aircraft first promoted theoretical studies of phenomena such as air turbulence, resistance and stream-lining and, once established, these physical principles were quickly applied to the refinement of the end product.

No one disputes that the origins of the nuclear power industry lie deeply buried in fundamental theories of the nature of matter and energy. This having been said, it is of little importance whether we take as 'the beginning', the implications of atomic 'number' residing within Mendeleev's periodic table of the elements (1869); the discovery of the electron in 1879 by J. J. Thomson, which showed that the atom is not indivisible; or the discovery of the neutron in 1932 by James Chadwick, which forced the conclusion that the same must be acknowledged even for the nucleus of the atom as well. Certainly the neutron had special and immediate significance, for it was a 'particle' without an electrical charge and could therefore be used to 'bombard' the nuclei of other atoms without suffering disturbances due to attraction or repulsion. In the hands of the Italian physicist Enrico Fermi – who by this time had fled from Mussolini to settle in Chicago – the neutron became a powerful tool with which the alchemist's dream of the transmutation of elements could be at last fulfilled. By bombarding the heaviest of the 'natural' elements, uranium, Fermi produced a man-made element, plutonium. In so doing he achieved, in the words of the commemorating plaque, 'the first self-sustaining chain reaction and thereby initiated the controlled release of nuclear energy' (1942). Fermi's 'uranium-graphite pile' was to be the model on which were based the early commercial nuclear reactors. Britain's first power station at Calder Hall began generating electricity in 1956 (although it must be said that its prime purpose at that time was the production of plutonium for nuclear weapons). The basic principle of a 'thermal' reactor such as this is the 'fission', or breaking up, of atoms of the isotope uranium 235. When this happens an enormous amount of energy is released, together with certain 'fission products', among which are neutrons. These neutrons can, under carefully controlled conditions, cause the fission of other uranium 235 atoms and thus start 'a chain reaction' by which a continuous supply of energy is made available. The heat produced during the operation of the reactor is removed by circulating water or carbon dioxide and used to convert water into steam for the generation of electricity.

In the modern world, the design and construction of aircraft (including spacecraft) and nuclear power stations is undertaken within enormous, research-intensive, industries commanding vast budgets and employing many thousands of highly trained personnel. These individuals are, in the broadest and most realistic sense, all working in applied science; only an arbitrary distinction could separate them into scientists on the one hand and technologists on the other.

What is clear about such individuals is that they are subject to motives and pressures which are very different from those of the relatively protected world of the universities. Not only are the traditional distinctions between science and technology, basic and applied science, broken down, but even that between established and specific academic disciplines may be unclear in the pervading atmosphere of inter-disciplinary collaboration for the good of the firm. How does the difference in work environment affect attitudes to work? Does the applied scientist feel a commitment to norms that are different from those of the idealized ethos of science, or of a particular research network? We still know relatively little about such matters, although the subject is an important one for sociology of (typical twentieth-century) science and it is well worth our spending a little time on what has been found to date.

Because the background and training of the modern industrial scientist is usually the same as that of the pure scientist, a number of American sociologists in the Mertonian tradition have assumed that he is the victim of conflict between two opposing 'cultures', those of pure science and of management respectively. According to this view the industrial scientist hardly knows which way to turn, for his own period of socialization during his university training has taught him to identify with the cosmopolitan ethos of pure science, with its ultimate goal of expanding knowledge, whereas the realities of his employment enforce loyalty on a much more local level to the practical objectives of his company and to 'ideals' such as financial soundness and a non-egalitarian, hierarchical organization. The stresses and strains not uncommon in the administration of the industrial laboratory are thus

Table 2 *Dimensions of work satisfaction: comparable indices for scientists and technologists in industrial research and universities*

	Scientists		Technologists	
	University	Industry	University	Industry
A Salary	55 (31)	77 (50)	81 (40)	76 (52)
B Quantity and quality of assisting personnel	79 (47)	78 (38)	65 (40)	63 (39)
C Amount of free time available for private research	90 (74)	38 (49)	73 (66)	38 (65)
D Opportunity for gaining experience in administration	20 (64)	63 (37)	32 (64)	55 (41)
E Prestige of this department in the scientific/technical world	60 (51)	43 (50)	67 (37)	51 (60)
F Prospects for promotion up a research career ladder	60 (49)	75 (44)	70 (38)	91 (54)
G Extent my qualifications and experience are fully utilized	81 (84)	94 (47)	95 (70)	85 (65)
H Opportunity to pursue basic research in my field	90 (83)	33 (60)	70 (85)	49 (65)
I Freedom to choose my own research projects	88 (96)	51 (49)	80 (85)	54 (56)
J Degree of freedom I have to manage my own work	96 (93)	91 (63)	88 (93)	90 (79)
K Opportunity to attend scientific/technical meetings/conferences	72 (60)	65 (51)	70 (45)	60 (67)
L Opportunity to work with highly reputed technologists/scientists	63 (77)	44 (49)	72 (38)	50 (61)
Number interviewed	50	118	40	75

Source: Modified from N. D. Ellis, 'The Occupation of Science', *Technology and Society* 5 (1969), p. 40.

explained in these terms and the industrial scientist is, at least by implication, seen as a frustrated would-be academic.

The limitations of this view have been highlighted by an empirical survey conducted in Britain by N. D. Ellis, in which a good cross-section of academic and industrial scientists and technologists (chiefly engineers) were asked directly what *they* thought of their conditions of work. Table 2 summarizes the results. The respondents were required to answer two questions about each of the listed conditions (A–L); firstly, 'how important do you feel that item is for your overall work satisfaction?' and secondly, 'how satisfied are you at present with each condition?'. The open figures in the table list the answers, within a range of 0–100, to the first question (i.e. degree of importance), and those in the brackets, the answers to the second (i.e. degree of satisfaction). The table contains a considerable quantity of information and is well worth studying. Thus the high scores returned by the academics in conditions C, E, H, I, and L, appear to confirm their attachment to the ethos of pure science. In contrast, the low scores of the industrial scientists suggest that, whatever their views immediately upon leaving university, they felt no such attachment, being significantly more pragmatic in orientation; the majority happily accepted the logic of industrial research and many were, indeed, even unaware that pure science might have a moral ethos of any kind. For condition D, the opportunity to gain administrative experience, there was a notably low score for the academics but reasonbly high scores for the industrialists; the role of administrator was evidently held in higher esteem by the latter who saw it as a more obvious path to promotion and as a means of achieving a broader perspective of, and greater relevance for, their technical work.

A further difference in attitude to administration emerged from Ellis's study; this was one between the scientists and technologists working within industry. The pure science graduate entering industry for the first time is often ill-equipped to cope with the unfamiliar conditions, particularly by comparison with the technology graduate whose training has prepared him better for the industrial environment. As a result, the scientist feels under-used (see condition G), finding that he is little valued for the science-specialist that he thought he was, and at the same time having difficulty adjusting to his new role as trainee technologist. By comparison again, the technology graduate readily identifies with his clearly designated position in industry and scores consistently higher than the scientist on conditions indicating commitment to his original specialism (E, F, H, I, and L). Partly as a consequence of his apparent insecurity, the scientist in industry is keener than the technologist to escape from the laboratory into other departments of the company where he might find a managerial role.

In summary then, Ellis's survey did not support the theoretical contention that any disenchantment on the part of scientists within industry is the result of their attachment to the so-called ethos of pure science. It seems that they mostly hold a favourable attitude towards industrial research and explain their marginal position – that is, as those who could be trusted only on technical matters, but not with the intricate subtleties of commercial policy – by reference to an educational tradition which has systematically under-emphasized the positive attractions of careers in technology.

Innovation

It has been argued that another fruit of this educational system is the relatively sharp distinction drawn by most people between the scientific and the technological-business communities. This separation is said to underlie the much-lamented absence of close links between pure and applied science. The importance of this problem is its connection with the question of how industrial or wealth-producing innovations actually arise. Do discoveries made in the course of fundamental 'curiosity-orientated' research lead unexpectedly to useful applications? Or is some specified 'goal' necessary in order to give direction to the basic research required to reach it? These two approaches are sometimes conveniently labelled respectively the 'discovery-push' and the 'need-pull' models of technological innovation; either accumulated knowledge pushes its way towards application, or perceived demand pulls out the appropriate research. In the book *Wealth from Knowledge* (1972) by J. Langrish and others, these two extreme positions are quoted and then sub-divided, not in any sense as definitive schemes, but rather as a basis for showing that in practice 'very few [innovations] fit any one of the above models.' A classic statement of the discovery-push models – intended presumably as a justification of pure research – is that by P. M. S. Blackett when President of the Royal Society: 'In a simplified schematic form, successful technological innovation can be envisaged as consisting of a sequence of related steps: pure science, applied science, invention, development, prototype construction, production, marketing, sales and profit.' The opposing view is expressed by J. H. Hollomon of the United States Department of Commerce: 'The sequence – perceived need, invention, innovation (limited by political, social or economic factors) and diffusion or adaptation (determined by the organizational character and incentive of industry) – is the one most often met with in the regular civilian economy.' The critical qualification of Langrish and his colleagues is important because it emphasizes that these alternative schemes are, of necessity, gross over-simplifications of what actually happens. To identify closely with either would be to fall into the trap of mistaking the model for the real thing, and few observers now believe that such linear schemes can do more than serve as very rough guides. When cautiously analysed in this way, several investigations have suggested in fact that need-pull is two or three times as effective as discovery-push, although it is acknowledged that the larger technological changes often tend to be of the latter type.

Most studies to date have shown that the conditions which promote technological innovation spread far wider than scientific or technical virtuosity. At least as much attention must be paid to external factors as to those of the laboratory itself. For example, the report entitled *Technological Innovation in Britain,* published in 1968 by the Central Advisory Council for Science and Technology, found five factors to be of outstanding importance. These were: the direct linkage of research and development activities with other functions of the firm; the planning of programmes of innovation on the basis of market research; the presence of a management that is commercially-orientated as well as technically efficient; the achievement of short lead-times between the beginning of research and the marketing of the new product; and finally, a realistic relationship between the product's launching costs and its production capacity and market size. In a word, the emphasis is heavily on the commercial orientation of the whole research and development process, which seems to be a clear advocacy of 'need-pull'. (This conclusion was the more striking for the fact that the report claimed basic research to be 'the fount of all new knowledge'.)

One of the general recommendations of the Advisory Council's publication was that some scientists and technologists should be deployed in stages of innovation other than those of primary research and development, for example in production or marketing. The sociologist Joseph Ben-David, says in further support of the general advantages of need-pull, that 'the link from economic and technological problems to fundamental research is more predictable than from fundamental research to economically useful technological innovation.' Accordingly, 'the optimal way to increase the uses of science is . . . not to select projects according to their supposed promise of applicability, but to increase the motivation and the opportunities to find uses for science, and to find practical problems which can stimulate research.' This will be best achieved by widespread and vigorous interaction between the ideas and problems of the laboratory and the world of management, an interaction which Ben-David sees as the function of 'entrepreneurs' who can exploit the theoretical advances of the scientists and bring them to the attention of the technologists and managers who will recognize their practical potential. This is again to acknowledge the stimulating influence of a diversity of roles and in this sense there is no difference between an effective utilitarian science policy and a policy for pure science such as that set out by Mulkay.

Many of these recommendations are clearly exemplified by the ways in which technological innovation is stimulated and assessed in a large multinational company such as the IBM Corporation. Its Vice-President and Chief Scientist, Lewis Branscomb, in first comparing the evaluation procedures of IBM with those of the United States National Science Foundation in allocating grants for academic research, makes the distinction between instrinsic and extrinsic factors in the perceived value of research. The former can be judged only by conventional peer review, while the latter must also involve the reactions of users and beneficiaries (including those making funds available). But Branscomb is anxious to blur any supposed analogy with 'basic' and 'applied' research, seeing this distinction as misleading and often a function merely of the way in which work is presented or 'packaged'.

At IBM (Branscomb was speaking in 1980) Research and Development expenditure is about 6.5 per cent of sales turnover, and some 15–20 per cent of this is of a clear 'research' nature. Although the 'quality' of the research output *per se* is naturally monitored as best it can be, this is only one of many interrelated elements that influence research decisions. Relentless scrutiny of all activities is, in fact, the order of the day. Thus there is constant 'measurement' of how effectively the research output is used in development initiatives, which provides an indirect assessment of the relevance of the fundamental work itself; evaluation of team leaders and other key staff (in terms of their personal attributes rather than of the reports they write) in order to see how creatively they interact and how they may best be developed and deployed; ceaseless pressure for financial efficiency; and an on-going vigilance concerning the 'scientific and technological image' of projects that gain support without the expectation of quick economic return. The last activity perhaps best represents the fundamental *raison d'être* for the Research Division; its function is to ensure that by its strategic results the Corporation will remain viably in business in the future.

A striking feature of IBM's 'value-management' is the trust that it places in the judgements of those who closely supervise the conduct of research work. It is these individuals – not some remote committee – who control the flow of funds to their particular work, subject of course to openly-assessed competition with other projects, and to the prying activities of a group of senior technical experts 'who have a particular charter to put their noses in places where they are not invited'. These individuals do not hesitate to report unfavourably on projects in which they have no confidence.

The main thrust of any company in the ruthlessly competitive world of high-technology must be concern with planning for the future, not evaluation after the fact. At IBM this is facilitated by a decision process that can direct – on the judgement of the Chief Executive – the total resource available to any selected priority. If this proves to be misconceived however, it can also be terminated quickly – an absolutely essential provision in an enterprise where hundreds of millions of dollars may be spent on just tooling-up for a new manufacturing process. The real point here is that resources must be aimed primarily, not at avoiding weakness, but at capitalizing on success; if the best innovations are not exploited efficiently the company will have lost a major investment opportunity.

IBM's Research Division is expected to justify every project, both technically and 'philosophically', so that background assumptions can be challenged in open debate. Moreover, its work is examined by the development divisions with a view to support or censure, as well as by a Corporate Technical Committee whose wider role is to estimate the state-of-readiness of the company in the face of international competition. Finally, there is a Science Advisory Committee composed of distinguished scientists from outside, whose task it is to evalute the laboratory research both for its scientific merit and its extrinsic value to the Corporation.

There is great contrast between the operation of a commercial company and that of a government agency when it comes to evaluating the quality of research in terms of innovations that are technically feasible, economically viable, and socially 'needed'. Few of the practices of general policy analysis, formulated and refined in other spheres during the last twenty-five years, have penetrated to the special area of science and technology policy, even during a period in which the science and technology enterprise itself has been transformed. There are as yet no rigorous standards on how the debate should be conducted, very few established rules of guidance on what constitutes legitimate analysis or on what basis advice might be provided to decision-makers, and only a handful of relevant cost-benefit calculations. With the partial exception of nuclear policy, over which there has been formal analysis of long-term consequences – but a great deal of disagreement – reliance is still placed in the 'well-informed' but largely informal and intuitive judgements of senior scientists and engineers. These tend to take the form of *ad hoc* justifications of a vague but superficially plausible kind, such as that a given research programme is necessary 'in the national interest', for 'preserving technological leadership' or 'improving educational excellence'. The assumption seems to be that one, at least, of these justifications must surely 'score'.

Among the problems faced by any attempt at science and technology analysis are the decentralized character of the whole enterprise, and the undoubted fact that we still know too little about it. If, for example, we return to the question of education, there is wide informal consensus, but still relatively little hard evidence, that an increasing flow of technical information between universities and industry would improve the rate of innovation. And there is confusion on science education more generally, for while the received wisdom defends it on utilitarian grounds (economic growth, international trade, etc.), some countries have grown and prospered in the absence of anything that could be considered high educational attainment. In these cases, education has not caused growth, but growth has made education feasible. Given that technology may *decrease* the need for high levels of scientific knowledge, it would perhaps be more appropriate to seek broader justifications in terms of human attributes such as creativity, flexibility and tolerance, which are desirable for life in a rapidly changing society. In this case, science education would have to compete (even in the context of efficient technological innovation) with alternative claims on time, energy, and resources.

POLITICAL AND ECONOMIC ASPECTS OF RESEARCH AND DEVELOPMENT

In the section on Technological Science above we discussed the aircraft and nuclear power industries as representative of the contemporary interdependence of science and technology. But any such massive enterprise is bound to have implications which extend far beyond the realm of theory and technique, and it is in just this connection that we have come to speak of the 'imperatives of technology' in modern society. The originator of this expression was the American economist J. K. Galbraith. In his seminal study, *The New Industrial State* (1966), Galbraith argues that it is the demands of technology which, more than any other factor, determine the major movements in economics, science, and society in general. In the capitalist countries, it is only the huge business corporations that have sufficient financial capacity to support the essential, but risky, technological research and development programmes which alone can promote change. Galbraith's general message has been widely taken up. Christopher Freeman, for example, asserts that

> for any given technique of production, transport or distribution, there are long-run limitations on the growth of productivity, which are technologically determined . . . Without technological innovation, economic progress would cease in the long run, and in this sense we are justified in regarding it as primary.

Galbraith himself attacks what he sees as the outmoded image of science still held by many people, an image that may have fitted the nineteenth century when it was 'the product of the individual efforts of men of genius', but which is quite inappropriate for modern science which is 'a highly organized new profession closely linked with industry and government.' Its success has been achieved by 'taking ordinary men, informing them narrowly and deeply and then, through appropriate organization, arranging to have their knowledge combined with that of other specialized but equally ordinary men.'

The links between science and government stressed by Galbraith and others have become increasingly important during the twentieth century. The very basis of the modern 'mixed' economy is a major involvement by governments in the activities of research and development. During the last fifty years there has been a ten-fold rise in overall government spending, which in most capitalist countries is now equivalent to about a quarter of the gross national product. Although the giant private corporations are able to undertake substantial research and development activities of their own, if they are to remain commercially viable in the long term they must also be in a position to exert considerable control over their markets (by advertising, market research and so on). Where this is impossible, or where the products being developed have one market outlet only (usually the government), there is no alternative to making public funds available to support the necessary research. This is why the design and manufacture of items for military and other strategic applications is so extensively supported by government finance, even though the work itself is often carried out in the laboratories and workshops of private industry. Thus in the United States, which since the Second World War has borne the brunt of Western defence expenditure, more than half of total government spending in the 1960s could be accounted for by military-related projects. This was some five times greater as a proportion than during the 1930s. Throughout the 1970s, in the wake of the American 'victory' in the race to the moon, the figure declined again, for example to about 35 per cent in 1973, but it remains enormous in absolute terms and in the 1980s is showing signs of rising once more as a result of policies of the Reagan administration.

The Military-Industrial Complex

The inevitability of government involvement in scientific and technological research concerned with national defense led, during the peak years of the 1960s, to the rise of a phenomenon known as the military-industrial complex. This has been studied most extensively in the USA although this should not be allowed to give the impression that similar arrangements do not operate elsewhere (including China and the Soviet Union). Close relations between the latest technological expertise and the requirements of the military, have always been a feature of a nation's preparation for war. As Freeman puts it, in the modern world 'the scale and complexity of . . . technology have been carried to extreme limits in research, design and development for military aircraft, missiles and nuclear weapons.' It is this enormous size and economic power that excites special interest, for despite the acknowledgement that in principle any enterprise equivalent in magnitude and technological sophistication could suffice as a substitute for high military-related spending in maintaining the health of the capitalist economy, in practice there is no sign of a comparable alternative on the horizon.

This extraordinary dependence on the so-called permanent arms economy has had several important consequences for science and technology. One that has aroused considerable alarm is the way in which the long-cherished independence of the universities to perform 'pure' research has been eroded by the progressive inclusion of academic science within the arms of the military-industrial complex. Significant proportions of the overall research budgets of some American universities are provided by the US Department of Defense, which functions as the 'customer', 'contracting out' work to specified laboratories. In common with military spending generally, this state of affairs reached a peak in the 1960s, but it must be said that since that time it has met with growing opposition and an attempt to free academic research once more from any such dependence.

Of greater quantitative importance was the tendency for government funds to be concentrated in a relatively small number of contracting companies. For example, in 1968, some two-thirds of US Department of Defense research expenditure was accounted for by 100 contractors (which at that time included two well-known universities), out of a total of over 22,000. This led to an alarming dependence of such firms on government (military-related) funding, and a so-called locked-in relationship arose in which such research became, for some, their sole *raison d'être*. Once established in this way, the existence of dependent companies – mere government satellites – has tended to be justified by the political 'need' for research into ever more sophisticated weapons systems. When the stage is reached at which some 10 per cent of the American labour force including, of course, large numbers of scientists and engineers, is supported by government military budgets, it hardly needs the 'Soviet threat' (or its equivalent in other countries) to ensure that weapons research continues on a huge scale.

It is obvious that this situation raises political and economic questions of a highly contentious variety. (Some of the specifically ethical issues are considered in the next chapter.) Suffice it to say that there are considerable differences of opinion as to the nature of the relations between the technological society and the escalating 'arms race'. To some, a massive weapons industry represents the only public spending component of the capitalist economy capable of regulating demand and avoiding catastrophic cycles of inflation and unemployment. To others, the arms race is simply the inevitable consequence of the international rivalry between the Big Powers, or just another example of the 'imperatives of technology', which determine military developments instead of responding to them. Which ever may be true, there is little reason for complacency and much for disquiet.

Costs and Benefits

In this section we can be concerned with only the barest outlines of the assessment of science and technology in economic terms. (This is a massive subject and readers with a special interest are referred to the bibliography for this chapter.) The notions of 'cost' and 'benefit' are also confined here to their limited, economic, aspects.

Since the Second World War, research and development expenditure in the industrial countries has grown several times faster than the respective gross national products. In the West, the leading nation in this respect has always been the USA, where in recent years between 2 and 3 per cent of Gross Domestic Product has been spent in this way. Among Western European countries, and in Japan, the figure is somewhat less (table 3). When these figures are broken down into 'policy objectives', a rough idea is gained of national priorities. From table 4 it can be seen that by far the top priority in Western countries, with a major post-War defence commitment, has been in the category 'National security and prestige', which groups together overt military

Table 3 *Resources devoted to research and development in OECD countries as percentage of GDP*

	1963/4	1971	1979
USA	2.7	2.5	2.4
UK	2.3	2.3	2.2
France	1.7	1.8	1.8
Germany	1.4	2.1	2.3
Italy	0.6[a]	0.9[a]	0.8
Japan	1.3	1.6	2.0

[a]Percentage of GNP.
Sources: *OECD Statistics*, Paris, 1975 and 1981 (simplified)

Table 4 *Percentage share of government research and development expenditure (simplified)*

		Security and prestige	Economic, agricultural, etc.	Health, welfare	Other
USA	1960	89	3	7	1
	1970	78	7	11	4
	1980	74	3	19	4
UK	1960	80	11	2	8
	1970	49	22	1	28
	1980	61	10	6	21
France	1960	69	8	1	22
	1970	51	18	2	29
	1980	50	13	14	23
Germany	1970	35	9	7	49
	1980	29	12	16	43
Italy	1980	32	22	11	35
Japan	1960	13	33	3	51
	1970	10	23	4	63

Sources: *OECD Statistics,* Paris, 1971, 1975 and 1981

spending with investment and nuclear programmes. However, between 1960 and 1980, there has been in most countries a decline in expenditure when it is expressed in this way. In Germany and Japan, with small defence budgets, spending on 'other' categories (including the 'advancement of knowledge') has been exceptionally high. The 'economic miracle' in the latter country especially is arguably a reflection of the substantial expenditure directed specifically to economically oriented research.

Again among Western countries (and Japan), there are some differences in the proportions of total research and development expenditure that are attributable to funds from governments and private industries respectively, although even more marked similarities. Table 5 shows, in rough terms, which of three sectors, government, private industry or 'other' (institutions of higher education, non-profit-making foundations etc.) provided money for research and development, and which sector spent it. Overall it may be seen that together, government and industry provide the vast majority of finance, and that industry performs most of the work. Whereas private firms invest their own money in research almost exclusively because they believe that in the long term there will be a return in the shape of profit-making

Table 5 *Research and development expenditure by different agencies in OECD countries as a percentage of total (simplified)*

		Source of funds			Sector of performance		
		Government	Industry	Other	Government	Industry	Other
USA	1969	57	38	5	17	70	13
	1979	49	46	5	14	68	18
UK	1968	51	45	4	25	67	8
	1979	41	43	16	21	64	15
France	1969	64	30	6	26	58	16
	1979	56	44	0	—	59	—
Germany	1969	42	58	0	16	65	19
	1979	47	50	3	17	65	18
Italy	1969	41	50	9	25	55	20
	1979	43	—	—	23	56	21
Japan	1969	28	60	12	11	61	28
	1979	17	59	24	12	58	30

Sources: *UNESCO Statistical Year Book,* New York, 1971 and *OECD Statistics,* Paris, 1981

innovations, governments are concerned with the profit motive in addition to strategic and political factors. Thus it is not surprising that support for non-market research, such as that in defence, health, agriculture and transport, is often provided by governments for no other reason than that it is judged to be socially necessary (i.e. in the absence of adequate private initiative). Support for 'fundamental' research (often carried out in the universities) is also provided by governments because it is recognized that really long-term and uncertain projects cannot be financed privately; any profitable outcome from these is viewed as an unexpected bonus.

Having now set the general scene in terms of 'costs', we can briefly address the question of greatest interest: are these very substantial research expenditures justified by economic benefits? Overall and in the longer term, it is not seriously doubted that research and development is profitable. From the viewpoint of the individual corporation, expenditure on research and development is a form of investment, and just like any other investment, it is made in the expectation of future benefit. Most econometric studies of growth (say, in per capita income) in the industrialized countries identify 'technical progress' as the single most important factor. If technical progress is manifested in innovation, and if economic growth is reflected in rising standards of living, the link between research and development on the one hand and consumer benefits on the other might seem to have been made. Yet except in some specific cases, a direct causal link remains hard to demonstrate, for the relations between research expenditure and rate of growth are by no means simple. A notorious analysis, based on national figures for research and development expenditures and subsequent rates of growth, and published by B. R. Williams in 1967, appeared to show an *inverse* relation between the two! Thus during the period 1950–59, the USA and Britain spent some three times as much on research, expressed as a percentage of GNP, as did Germany and Japan, yet grew far more slowly (as measured after a five-year lag to allow the 'benefits' to work through). Since no one would pretend that economic growth could be achieved directly by *reducing* research expenditure, the explanation for the relationship must lie elsewhere. As may be expected from what has been said above, it is chiefly to be found in the far greater military (i.e. non-economic) government expenditures by the USA and Britain. But there is also the fact that the useful results of other people's research are often readily available through the purchase of patents, which can rapidly diffuse new technologies throughout the world.

At any rate, as Williams says, the graph shows clearly enough that high research and development spending alone is not a sufficient condition for economic growth (despite its apparently being a necessary one). Even when expenditure is confined to research and development in the so-called economically motivated sectors, there is generally no direct correlation with growth rates. Clearly therefore, no simple linear model of the type:

research (invention) – development – innovation – production – sales – increased profits – increased investment – higher GNP

can hope to account for the complex networks of interacting factors at work. Innovative activity is by no means the only input essential for profits and growth. It is much more a question of how successfully a country *uses* its scientific and technological advances for commercial purposes, and this will depend upon qualitative judgements as to the best balance between a host of operative market factors. Nor is it enough to consider national conditions alone, because the competitive market for high-technology products is world-wide.

This international picture is complicated to say the least. The various attempts by economists to measure the contribution made by science and technology to economic growth have had only limited success, although they do appear to provide general confirmation for the almost universal 'belief' in the efficacy of research activity, by showing that growth cannot be explained *without* its contribution. A difficult problem has been that the great variations in quality and quantity of research between nations have rendered many of the comparisons which have been made largely meaningless. Thus while massive support by the American, British and French governments for the high technologies of aerospace and nuclear reactors has provided little evidence of resultant economic advantage, the minimal government involvement in a small country such as the Netherlands, which specializes largely in sophisticated electronics, has certainly not been reflected in economic stagnation. Germany and Japan, with modest post-war military spending, have enjoyed rapid and sustained economic growth while Britain, with the expense of her 'independent' nuclear deterrent, has fallen gradually behind. It is arguable that Britain's relative weakness in the twentieth century is something of a repetition of her experience in the nineteenth when, after early industrial leadership, she was quickly overtaken by Germany and the United States owing predominantly to a failure to invest in new and improved technology. Then, as in more recent times, the 'soft' markets built up in the Empire and Commonwealth, seduced the first industrial nation into a sense of security and invulnerability that was altogether unjustified. Some economists speculate that a similar experience awaits the USA. So long as American companies continue to seek profits by investing in easier foreign markets, and so long as economically oriented industrial development at home is retarded by the disproportionate investment of funds and skilled personnel into military-related research, the country risks being overtaken technologically by Germany, Japan and others.

Be that as it may, the unfortunate reality seems to remain that the transfer of a significant proportion of the research expenditure of governments – all governments – towards problems of a more socially valuable kind, awaits not just the availability of accurate methods of economic measurement and prediction, but a radical shift in the international political climate. In the meantime, research and development will continue to serve the whims of governments, and nations 'will get the science they deserve'.

7

Ethical Dimensions of Science

As a self-conscious part of society at large, the community of scientists can no longer opt out of the great moral issues of the day behind assertions such as that of Marie Curie (1867–1934) that 'science deals with things, not people'. Manifestly, in our own day science does deal with people, not only indirectly but directly too, and more and more there is the danger that it will deal with people *as if* they were only things. Furthermore, with the rapidly increasing concern about the depletion of non-renewable resources and the burgeoning 'ecology' movements, it is also no longer clear to everyone that 'things' – both animate and inanimate – do not in any case raise moral dilemmas of much the same kind as people do.

The greater part of this chapter will be concerned with what is called 'applied' ethics, that is, with the moral aspects of particular modern scientific practices. But to begin with it is necessary to construct a platform from which such questions can be considered. To do this we must look in an elementary way at the more general ethical relations between science and social values. Such relations pose special problems for the writer because he is likely to feel unsure as to whether he should attempt a genuinely impartial stance on issues of value or whether it is more realistic to advocate a personal, even a thoroughly partisan, position. In what follows, the emphasis has been – except where the contrary is made clear – on objective discussion, although by this time the reader will be well aware of the limitations of 'objectivity' and will know how to respond if he detects elements of personal bias with which he disagrees.

ETHICS, SCIENCE AND SOCIETY

The ancient Greeks generally made a distinction between natural philosophy and moral philosophy, the former being concerned with the things and events of the world, the latter with the conduct and aspirations of man. The natural world could be studied empirically and rationally, but moral questions were in a different category altogether. Ultimately, it seemed that they could be resolved by the individual only by an appeal to his deepest *feelings* of approval or disapproval, to his intuition of what was right and what was wrong. The distinction between those questions which were amenable to reason and those which were apparently not, was decisively re-emphasized by the Scots philosopher, David Hume, who argued that it is not, on the basis of what *is* done, logically possible to determine what *ought* to be done. Hume's argument itself had great influence and the work of the German philosopher Immanuel Kant (1724–1804) had the effect of further reinforcing it by identifying two kinds of mental activity, pure theoretical reason which governed the possibilities of human experience, and pure practical reason which governed moral action. Speaking simply, we can now see that it was from this philosophical basis that the 'typical' early twentieth-century attitude towards science and ethics, such as that of Mme Curie, was derived. Science was descriptive and impersonal, concerned with the world 'out there', while ethics was prescriptive and personal, concerned only to guide us towards right action.

It took the impact of World War to show that what it was right to do within the closed and detached world of science, was not necessarily right in the wider context of society at large. Thus it was soon realized that there might be a clash between the impersonal interests of science and the more human interests of society, or at very least that there must necessarily be *inter*action. Society, it could be readily recognized, is a community held together predominantly by the acceptance of ethical and behavioural norms; a widespread rejection of these norms would quickly lead to disunity and collapse. But the scientific community *within* society at large is also governed and united by ethical norms – for example, the obligation to report findings truthfully – and it is only by the acceptance of the respective norms as they refer both to society and to science that agreement as to what constitutes social and scientific 'facts' is possible. (It is, indeed, mainly because the ethical norms of contemporary society are less widely accepted than are the norms of science, that there is so much more *dis*agreement about what 'facts' really are in the social sciences as compared with the natural sciences.)

When there is conflict between the norms of science and the norms of the wider society, problems occur of the kind with which we have become all too familiar during the last twenty years or so. Quite apart from the ethical dimensions of the activity of science itself which, according to Merton and others, regulate the behaviour of the scientist *qua* scientist, a clash between science and society enforces an examination by the individual of his respective loyalties to the *two* communities. Should he pursue the 'truth' as dictated by science no matter where it may lead, no matter what the consequences for society, or are there circumstances where his responsibilities as citizen must transcend his obligations as scientist? Is there something unique about the scientific enterprise which makes it a desirable model to be copied by society itself? Or would a fully scientific society drain out the warmth and beauty that might ideally be at the very heart of human experience?

'SCIENTISM': SCIENCE AS THE MODEL FOR SOCIETY

The answers to questions such as these can be formulated only within a particular view of science derived from a particular philosopohy. (This is one of the most important reasons why a modern student of science has much to gain from a grasp of the essential principles of the philosophy and sociology of science.) Those individuals who construe science as nothing less than 'truth institutionalized'; who see in it something akin to the Cartesian ideal of 'purity'; who broadly accept the moral norms set out in Merton's scientific ethos (see previous chapter); will, not surprisingly, see in the community of scientists something very special and very worthy of emulation. Unlike most human communities, the community of scientists may seem to be democratic, disinterested, tolerant and, above all, rational. It will appear uniquely successful at achieving organized stability, while retaining freedom for individuals in an association of mutual trust. With such an impressive array of characteristics, the institution of science will naturally seem the obvious model for the organization and ethical basis of the wider society, and even for a Utopian world community.

The high esteem enjoyed (at least until recently) by science in modern society is a reflection of the widespread, implicit adherence to views of this general kind. To append the word 'scientific' to one's evidence is to lend one's argument weight of a special kind; to associate it with some action is to suggest a specially notable reputability. The use of idealized scientific images as a source of authority in this way has become known as 'scientism', and the discourse using them is described as 'scientistic'. Scientistic writing is commonly concerned to define ethics as a function of science. A good example is the widely quoted article, 'A Scientific Approach to Ethics' (1957) by Anatol Rapoport. Rapoport sets out what he considers to be the ethical principles inherent in scientific practice. These principles are

> the conviction that there exists objective truth; that there exist rules of

evidence for discovering it; that, on the basis of this objective truth, unanimity is possible and desirable; and that unanimity must be achieved by independent arrivals at convictions – that is, by examination of evidence, not through coercion, personal argument, or appeal to authority.

For Rapoport, these principles are unique and they provide 'a peculiarly suitable basis for a more general system' of ethics.

Rapoport's scientism is founded upon an empiricist philosophy of science, one based on the belief that an objective truth lies, as it were, 'out there' waiting to be revealed by the rules of scientific method. Although there are many variations on this general theme, the position has been supported in broad measure, and in recent times, by such scientist-philosophers as Jacob Bronowski, Michael Polanyi and Jacques Monod. (The inclusion of such names as these, incidentally, should dispel any illusion that this attitude towards science and society is necessarily related directly to political outlook. Although people with an interest in these matters naturally tend to have their own views as to what the ideal society would be like, Bronowski was politically a socialist, and Polanyi politically a conservative.) In his highly influential book, *Chance and Necessity: An Essay on the Natural Philosophy of Modern Biology* (1971), Monod rejects the claim that objective truth ('what is') and the theory of human values ('what ought to be') are eternally opposed, on the ground that knowledge and ethics must inevitably be linked through action. Yet he does still wish to maintain a radical distinction, arguing that 'no discourse or action is to be considered meaningful, *authentic,* unless – or only insofar as – it makes explicit and preserves the distinction between the two categories it combines.' In positing the principle of objectivity as the condition of true knowledge, Monod recognizes that this in itself constitutes an ethical choice.

> In order to establish the *norm* for knowledge the objectivity principle defines a *value:* that value is objective knowledge itself. To assent to the principle of objectivity is, thus, to state the basic proposition of an ethical system: the ethic of knowledge . . . The ethic of knowledge does not impose itself on man; *on the contrary, it is he who imposes it on himself,* making it the *axiomatic* condition of authenticity for all discourse and action . . . . The ethic of knowledge that created the modern world is the only ethic compatible with it, the only one capable, once understood and accepted, of guiding its evolution.

Evolutionary and Ecological Ethics

For Monod the ethic enshrined in science is the only one fit to guide the future development of human society and the natural world; he articulates an uncompromisingly scientific picture of nature. In another scientific tradition, the study of nature itself is thought to reveal to man patterns which represent the true, objective and 'natural' basis for his own society. Despite the logical difficulty of extrapolating from a description of nature to a prescription for man (sometimes referred to, after the philosopher G. E. Moore, as the 'naturalistic fallacy'), it is argued that phenomena such as evolutionary 'progress' or ecological balance are somehow intrinsically 'good' for the human community since this community is, after all, an integral part of the natural world. Thus certain conditions of change or of stability in nature are held to be desirable models for human conduct, such that action in conformity to them is taken to be right action.

Evolutionary ethics was originally a nineteenth-century development which sought legitimation in the arguments of Darwin's *Origin of Species*. In the form of Social Darwinism it had considerable impact in Victorian Britain and in America. A much more sophisticated variety, lacking the somewhat sinister overtones of Spencer's original expression and extending the general ideas to 'sociogenetic' or 'psychosocial' evolution, was espoused in Britain from the 1940s, particularly by the biologists Conrad Waddington and Julian Huxley. The central theme has always been the concept of progress, in particular of course the progress of man. The general direction of this progress is, it is argued, discernible in nature, the direction is 'good . . . according to any realist definition of the concept' and man would be well advised to shape his aims and institutions in congruence with this natural trend.

Unfortunately for evolutionary ethics there was from the earliest days a difference of opinion between those such as Spencer himself who emphasized the 'goal' of Nature as the ultimate attainment of perfection, and for whom the means whereby this goal was achieved were unimportant, and those such as T. H. Huxley, who took diametrically the opposite view, seeing Nature as characteristically 'red in tooth and claw', a force to be actively opposed by man in the name of the ethical progress of his society. Likewise, there were always differences as to whether the central theme to be derived from organic evolution was that of competition between species or co-operation within species. Preference for one or the other tended to reflect prior views as to the motivating force of the 'ideal' society; competition as the recipe for progress by elimination of the less fit, or mutual aid as the guarantor of the safety and survival of all.

It is this question of survival which is the dominant concern of the much more recent school of ethics fathered by the ecologists and conservationists. But in the other respects too there are many parallels with evolutionary ethics. In ecological ethics however there is great emphasis on such concepts as the harmony, stability and delicate equilibrium of nature as a self-recycling energy system, and a clear advocacy as a model for human society of the closely integrated, richly interwoven and 'self-sufficient' pattern derived from an idealized image of nature. The maintenance of this equilibrium has been called 'the first law of the morality of Nature' and man is seen inevitably to be a part – by now the most important part – of the whole cycle. Special fears are expressed concerning the dangers of interfering with the equilibrium, in particular because we seldom understand fully the consequences of what we do.

The ecologists do not have it all their own way, of course, for there are also vociferous opponents of conventional ecological ethics who, like Thomas Huxley before them, stress the 'imperfections' and prodigal wastage of nature and who advocate upsetting the natural balance to man's supposed advantage. If necessary, this might even mean interfering genetically to eradicate the 'deficiencies' of natural selection. As with evolutionary ethics, there tend to be on both sides of the argument pre-existing notions of what *is* good for society, notions which to a significant degree determine the elements in nature that are held up for admiration or distrust.

SCIENCE IN THE SERVICE OF SOCIETY

In simple terms the doctrine which represents science as the model *for* society may be understood as a natural development of the *Method* set out by Descartes. Similarly, the opposite view of science as a co-operative venture concerned with the gradual accumulation of knowledge of benefit *to* society, may be described as the application of Bacon's *New Instrument*. In the Cartesian view it is science which, as it were, takes precedence; in the Baconian view, it is society. Although Baconian inductivism was rejected by the modern historian Thomas Kuhn (see chapter 4), the latter's philosophy, and in particular its notion of truth, is in some respects a basis for contemporary pragmatic science. Kuhn, it may be remembered, urges that the raw facts of science are inevitably accumulated within the context and under the guiding hand of theoretical structures which he calls paradigms. Such absolute principles as Truth or Trust are eschewed; being permitted only within the relativist frame of a given paradigm. Normal science is, for Kuhn, a somewhat humdrum activity akin to a game, which attempts to solve specific puzzles by particular rules, both the puzzles and the rules being defined by the paradigm to which the scientists in question adhere. A solution to a puzzle may represent a limited kind of truth, limited because it is merely a function of the particular rules followed; but the scientist does not search for, nor does he enquire about, any *absolute* truth for this is a concept beyond the confines of science. According to this view, the scientist is guided not so much by a community devoted to a refined and universal ethos of science, as by one which shares a common, but restricted, paradigm.

Some advocates of this general position go further, arguing not only that scientists fail to pursue the Truth as set out by the norms of a scientific ethos, but that the ethos itself is no more than an idealized conception bearing little resemblance to reality (see chapter 6). Not surprisingly, those who hold so profoundly critical a view of the relevance of the ethos for 'real' science generally have little to say in favour of its wider application as an ethical basis for society. For such critics science should serve not as the model, but rather as the servant of society, being directed specifically towards the solution of social problems. Consequently they answer in the negative Bacon's famous question: 'Is truth barren?' and recommend the planned application of science to 'endow the life of man with infinite commodities'. Even the ethical nobility of pure science is implicitly questioned, for is it not morally preferable to seek mastery over nature for the benefit of man rather than to investigate the mystery of nature as an end in itself?

Unfortunately, this fundamentally important question tends to founder upon the familiar problem of defining the Good. What *is* good for society is notoriously elusive, for it is dependent upon a host of personal, political and economic points of view. This, indeed, is why modern science so characteristically tends to serve ambivalent purposes. But in an age when even the purest of Cartesian science cannot escape dependence upon external institutional support – sometimes of an enormously high order, as in sub-atomic physics – the attempt has to be made to confront such difficult problems. Ethical questions as to what *should* be done are inextricably linked with economic questions as to what society can afford to do. In other words, choices have to be made.

Criteria for Scientific Choice

The realization that modern science must perforce choose between the competing claims of its different sectors seems very substantially to have pre-empted the long-running debate over the desirability and dangers of an overall, planned, policy for science. In Britain the two sides of the debate were for many years associated with the names of the crystallographer Desmond Bernal (a Marxist) and the chemist Michael Polanyi (a conservative). Bernal's classic book, *The Social Function of Science* (1939) presented the case for the planning of science in the context of the needs of society, stressing the requirement for adequate investment in science and the appropriate use of manpower. At the time it seemed a radical, even a threatening, thesis and it provoked a number of counter-attacks from the more traditional school of the academic liberals, among them Polanyi's book, *The Logic of Liberty* (1951). Polanyi was concerned not merely with what he saw as the danger to the individual scientist's freedom, but more fundamentally with the argument that any curtailment of this freedom in the name of organization or planning would necessarily paralyse the scientific enterprise itself. Thus he came to formulate his famous concept of the 'Republic of Science', an autonomous system which can advance only by the guidance of something akin to Adam Smith's 'invisible hand' – the combined result of all actions by the community of free individuals.

The event which undermined this controversy was, of course, the Second World War. This quickly compelled the realization that priorities had to be agreed as to the direction of research effort, for funds had to be allocated preferentially in the over-riding national interest. The enforced acceptance of this new situation by scientists during the war was carried over into the subsequent peace, so that the relevance of Polanyi's case came to be restricted to the diminishing sector of 'pure' science, as distinct from science as a whole. Accordingly, the question as to the planning of science became re-phrased. It was no longer doubted that science *should* be planned; it was rather a matter of deciding on what basis the planning should be carried out.

Probably the most influential basis for science policy has been that suggested in 1963 by the distinguished American physicist Alvin Weinberg, then Director of the Oak Ridge National Laboratory in Tennessee. Weinberg applied himself explicitly to the problem of how priorities could be established for the funding of scientific projects. He specified two types of criteria, those internal to a given field of science, and those external to it. The internal criteria are intended to answer two questions; first, Is the field ready for exploitation? and, second, Are the scientists in the field really competent? These questions can be answered, says Weinberg, only by scientists themselves, so they may be referred to panels of appropriate experts who will judge the quality of research activities and personnel.

The controversial element in Weinberg's analysis concerns the external criteria. By arguing that 'it is not tenable to base our judgement entirely on internal criteria', he is clearly advocating the subordination of purely scientific goals to social goals. Such a Baconian or utilitarian scheme has, in fact, long operated in most countries of the world, and the importance of Weinberg's paper is that it treated the issues overtly.

He identified three external criteria, namely, technological merit, social merit and scientific merit. The first is relatively straightforward, for once some technological goal is deemed worthwile we are obliged to finance the scientific research necessary to achieve it. A topical example would be the medical technology of organ transplantation. If society has decided that this is desirable, then it has no choice but to support the basic research in immunology which alone can hope to overcome the problem of tissue rejection. The argument here is not intended to imply that the technological relevance of basic research is always apparent or that the fundamental research most likely to solve technological problems will necessarily emerge as a result of their recognition, but merely that the solutions to problems associated with a desired technology may be expected to come more readily from work geared specifically to them, than randomly from work that is otherwise unrelated.

Here, of course, the use of the word 'desired' indicates the inevitable entanglement of technological and social criteria. Weinberg acknowledges the grave problems that arise from any attempt to judge the relevance of scientific and technological work to human welfare and human values, for who is to define the values of society? Perhaps wisely, however, he offers little guidance on how the dilemma may be resolved. Even with 'rather uncontroversial' values, such as 'adequate defence, or more food, or less sickness' he is aware that it is by no means easy to determine in advance if some particular research programme will indeed further their pursuit. Two points are perhaps worth making. The first is that in the context of an enterprise as characteristically international as science, it must surely be hoped that concepts such as defence, food and sickness can be discussed in global terms; within a utilitarian ethic, not to do so would be to beg the really important human questions. The second point is that, even within the more limited boundaries of national science policy, a value such as 'less sickness' is also highly complex. The case of heart transplants, for example, raises many delicate issues. Are the lives 'saved' by such drastic measures of equivalent 'value' to the lives of new-born babies lost through the provision of inadequate, but relatively simple, resuscitation facilities? With improved success-rates, is the (glamorous) technique of transplantation likely to be financed at the expense of (mundane) measures to alter life-styles or environments? And is the sophisticated transplant technology becoming an end in itself which could lead to more bizarre transplantations?

Weinberg's criterion of social merit clearly involves value judgements of the most difficult kind, external to science altogether. In connection with his final criterion, that of scientific merit, it is important to emphasize that considerations are involved, not outside of science itself, but only outside of the field in question. If judged solely from *within* its own field, a science is likely, say Weinberg, to become 'baroque', that is, trivial and self-fulfilling. Any field of science is, in reality, embedded in many related fields and the suggestion is that a field's vitality be measured by its involvement with these others as an integral part of a universal, overall science. 'That field has the most scientific merit which contributes most heavily to and illuminates most brightly its neighbouring scientific disciplines.' This criterion is original and, together with the others, particularly interesting to apply to specific examples. It also seems to have independent support in the sociological evidence showing that innovation is promoted by 'cross-fertilization' (see chapter 6).

Weinberg himself assesses several scientific and technical fields on the basis of his own criteria. He concludes that molecular biology should be supported with maximum generosity for, he says, it ranks highly on all criteria, notably in its impact on related sciences (cytology, genetics, microbiology) and in its basic support for medicine. In contrast, he rates high-energy physics – at present the most prestigious branch – poorly on all external criteria, identifying the enormous expenditure as its Achilles' heel – in so doing he implies that a ranking which is low matters less when costs are also low. On the question of the social sciences Weinberg concludes that there are strong reasons for support because of their good scientific merit (they are all interrelated) and their obvious relevance to human affairs. Although they may lack a clear sense of direction, the social sciences have the great advantage of costing little.

THE SCIENTIFIC SOCIETY AS IT IS

In the every-day world, the ethical relations of science and society are not nearly as clear-cut as the rival, idealized, theories would wish. The scientistic (and perhaps Utopian) dream of a society dictated by the purest ethics of science seems ever further from realization, while the regulation of science exclusively by social goals has still not gone nearly far enough for the radicals. In fact, a shifting, uneasy relationship prevails at the present time, although a pragmatic orientation without any absolute objective ideal appears increasingly to dominate. In the minds of most informed observers, the ethos of pure science now describes the moral imperative of a very tiny and very unrepresentative part of the scientific enterprise. Mission-oriented Big Science – piecemeal and problem-solving – is what really counts.

The failure to translate the images of scientism into social reality appears to have been the result of the inapplicability of the ethics of pure science to the human world beyond. By and large, the ethical ideals of science are objective, impersonal and international, whereas those of human society are subjective, interpersonal and national. A decision to define the first set as the more fundamental is itself a matter of moral judgement, and there has always been the haunting fear that a 'scientific' society based on these norms might have no place for such age-old human values as honour, self-respect, or simple trustworthy behaviour between individuals (no matter how highly these are supposed to be regarded by the scientific community itself). It might, furthermore, undervalue deep emotional relations which cannot be quantified, in its relentless pursuit of organization, efficiency and power. Thus despite the widespread respect for science, there is also considerable suspicion: a feeling that science may achieve miracles with natural phenomena but that it should not be allowed to extend too far into the sensitive affairs of individuals and of society.

We need not speculate more deeply on possible reasons for this suspicion, nor on the apparent incompatibiity between scientific and social ethics. Suffice it to say that a good deal of control over science has, in fact, been achieved, such that the most urgent question is no longer the desirability of the subservience of science to society or of society to science, but that of the human needs which science should be geared to satisfy. Here, however, we encounter some of the most delicate and emotive issues of all, for it is seldom possible to achieve a consensus on what the *real* human needs are. During the past generation the world has been dramatically transformed by the impact of science and technology, yet it is extremely doubtful if this transformation has occurred at the behest of human 'will'. On the contrary, it is widely held that the information-communication revolution wrought by the electronic and computer engineers occurred, and is continuing apace, because it was and is 'inevitable'. According to this view, science is not actually controlled by the wishes of humanity at all. It is no longer construed as a Baconian Instrument for fulfilling specified human needs, but is governed by something akin to Galbraith's 'imperatives of technology' (chapter 6). The essential modern message has become: 'What can be done, will be done', even if this means the transplanting of human heads (said already to be close to technical feasibility), the 'cloning' of a race of super-men (still a long way off) or the destruction of all life on earth (possible or probable according to your judgement). The tendency to impute a force of inevitability to the 'progress' that results from the application of a so-called 'neutral' scientific method (see later in this chapter), independently of scientific or social ethics, has always commanded the implicit support of a number of prominent scientists. Among these was the physicist and political liberal J. Robert Oppenheimer who reluctantly defended the development of the hydrogen bomb on the ground that it was 'technically sweet'. The whole alarming syndrome may be summed up in typically caustic fashion by the words of Bertrand Russell: 'Whatever folly man is capable of conceiving, man has always historically performed'.

. . . And as it could be

The allegiance of the scientist can no longer be to a methodology alone. A science that seeks purely scientific goals poses no problem when it operates on the level of a cultural activity, but when it has reached the massive scale and potential of late twentieth-century science, and in particular when it operates within a society which lacks a universal ideology of its own, it presents dangers of the most awesome kind. The ethical imperatives that operated reasonably enough in the nineteenth century seem manifestly inadequate in the present day, such that science and social policy no longer have a choice but to come together.

But what would be the nature of any close relationship and how could it be achieved? The reader will be in no doubt that there are many possible answers to these questions, each reflecting attitudes which in part at least are normative or 'merely personal'. One persuasive approach argues that the most rational way of minimizing the tension which at present exists between science and society is to expedite a new degree of mutual comprehension between them. Implicit in this approach tends to be the idea that science needs to be 'humanized', that is, made more relevant to human needs and aspirations. Accordingly, it is said, the methods and objectives of science must be endowed with compassion, and consciously directed to the reduction of suffering and the enhancement of the 'human spirit'. In support of this general attitude – if it needs empirical support – are stressed the results of innumerable social studies which seem to indicate that while human happiness, notably the sense of purpose, commitment and fulfilment, *is* of course related to the satisfaction of basic needs such as food, shelter and hygiene, it is not an obvious function simply of material wealth. A scientific-technological society which provides a rich variety of luxury consumer goods at the expense, let us say, of a feeling of brotherhood and community, is evidently not a society which serves the good of man as well as it might.

When the concept of the human society is made international and when it is realized that some two-thirds of its individual members suffer malnourishment or under-nourishment, many dying of diseases long since eradicable by medicine, any ideal of universalism in science or society is seen to be well short of realization in practice. Furthermore, when the United States alone is acknowledged to spend each year more than two thousand times as much on alcohol and tobacco as the sum estimated by the World Health Organization to be necessary (yet not forthcoming) for effective remedial measures on the problems of major Third World diseases, can the suggestion that science should give priority to those problems of contemporary society which it can readily solve, any longer seem unreasonable? It would not seem so if science and society could be made to coalesce to the point where the problems and priorities of the one were a part of the problems and priorities of the other.

What many might regard as an ideal basis for the ethical relations between medical practitioners and their patients and, by implication, between science and society, has been outlined by the philosopher William May. After comparing systems based on more familiar forms, such as contract or philanthropy, May accepts the covenant 'as the most inclusive and satisfying model for framing questions of professional obligation'. Briefly, the covenantal ethic is based on the acceptance of bonds of mutual trust and mutual debt. For it to operate successfully, both science and society would therefore have to acknowledge that each has been the recipient of gifts from the other, most obviously in the form of financial support and of technological innovations respectively. In return for these gifts each must also accept the obligation to give to the other, not according to some rigid contract where reciprocal exchange is agreed in advance, but without calculating the return, according to a selfless code of conduct similar to that operating within a Mertonian ethos of pure science. Each party to the covenant would give that commodity which it best produces, thus gaining esteem and satisfaction for itself and providing for a particular need of the other. There is clearly a 'scientistic' – not to say Utopian – element in May's system, yet without the essential ideal of obligation and trust it is a great deal more difficult to envisage a truly humane basis for a science-based society.

THE 'NEUTRALITY' OF SCIENCE

A genuine Cartesian science – pure, disinterested and without ulterior motive – would by definition be neutral, in the sense of being morally and socially value-free. Such a science could be likened to any other 'harmless' cultural or aesthetic activity, for it would be little more than one among many ways of exploring man's multifarious relations with the universe. But we have already made clear in several connections that even if so pure a science is anywhere still practised, it is not remotely representative of modern science as a whole. Questions concerning the possible neutrality of science must therefore consider applied science, and perhaps technology, as it actually is in the modern world. Can it be that science in this broader sense is neutral as between good and evil? Does the individual scientist carry any special responsibility, *as a scientist,* for the work that he does or for the uses to which his work may be put by others? (It should perhaps be stressed at this point that a distinction is here being made between science-as-knowledge or science-as-knowledge-seeking on the one hand, and science-as-social-activity on the other. The value-free nature of the first two of these categories may be relatively simple to determine; in this section however we are concerned only with the final category.)

We might not be too surprised if the members of society keenest to subscribe to the notion of scientific neutrality were the scientists themselves. Many practising scientists, unaware or dismissive as they are of philosophy of science, cling to an idealized conception of their profession and propagate a view of 'scientific truth' which implies complete certainty, objectivity and detachment. Such a view may be held in the full knowledge that many kinds of science can be practised only by virtue of financial support provided by governments or industrial companies with goals which are frequently unclear, and almost always directed by political or economic interests. Incompatible as these two positions may seem, it yet remains true that the prestige and authority of science is such that they are widely accepted, more or less unthinkingly, by the public at large. To some degree at least, scientism has usurped the territory formerly held by religion.

Can the thesis of neutrality stand examination? The ends of modern science are now very largely ruled by the Baconian ideal of 'dominion over nature', that is, by a science policy which identifies knowledge with power. Furthermore the means of science – those means by which its problem-solving activity may be undertaken, as well as the kind of answers which may be offered – are now widely acknowledged to be dictated by the paradigms which currently prevail. But the relations suggested by a paradigm are themselves commonly influenced by social and political factors, so that the problems selected for investigation by the individual scientist, by the commercial company, or by a government department, will reflect value judgements as to what it is important to do. Even a belief that the pursuit of *any* knowledge is worthwhile requires an act of judgement, and certainly the decision that some kinds of knowledge (for example that connected with research into recombinant DNA) are better *not* pursued is deeply impregnated with personal values.

It has long been recognized that scientific knowledge, however 'pure' at the time of discovery can, in the hands of others, be a power for evil.

Is science, or are scientists, to be held responsible for the abuse of scientific discovery? Was Einstein, for instance, responsible to any degree for the atomic bomb which was the practical fulfilment of his purely theoretical insights as to the interconvertability of matter and energy?

The case for CS gas is a more recent, and perhaps a more controversial, example. Its development into a widely available harassing agent for crowd control and for wartime use can be traced back to an apparently innocuous research publication in the *Journal of the American Chemical Society* (1928), entitled: 'The Reactions of Alpha, Beta-Unsaturated Dinitriles' by B. B. Corson and R. W. Stoughton. Aspects of this paper were 'spotted' as having potential application, and the work necessary to produce an incapacitating gas for dispersing crowds, and ultimately for 'flushing-out' people from the caves and tunnels of Vietnam, was performed by Britain's Chemical Defence Experimental Establishment. According to Steven and Hilary Rose, who describe the whole incident, the finger of suspicion must be pointed very firmly at

> the War Office directive of 1956, which set forth (the) new research direction. Here, in the transition from a scientific paper to a technological concept, lies a key point in the application of science with respect to society, a point at which it is perfectly possible to identify responsibility and non-neutrality.

The contention here is therefore that the two chemists could not themselves be held responsible for later, unforeseen, developments of their own work. Responsibility must lie with the Ministry of Defence committee which sent the directive *without,* let it be noted, any public consultation. The Roses provide further parallel examples and claim that 'they are intrinsic to the current development of science', suggesting that it is scarcely accidental that the book describing the development of another weapon used widely in Vietnam and elsewhere, napalm, is called nothing else than *The Scientific Method* (by the well-known chemist L. F. Fieser).

The attitude of the Roses may seem extreme to some, yet their examples give much food for thought. They suggest that even the purest of science cannot (although the purest of scientists, perhaps, can) retreat behind the shield of alleged neutrality. Where the science is less than pure, which is the case with the overwhelming majority of all modern research programmes, any claim to neutrality would seem impossible to justify. Thus for many people applied science, in contra-distinction to 'Cartesian' science, is by definition not neutral; it is a social activity overtly geared to the discretional priorities of governments and industries. And because most so-called basic research is now performed in industrial laboratories (that is to say, it is 'oriented basic research' – see the previous chapter) it too cannot escape the dilemma of responsibility, such that it may no longer be feasible to separate basic research from its application for purposes of moral accountability.

SOME ETHICAL DILEMMAS

The above conclusion seems to be reinforced by the host of practical problems that have recently come to public attention. Before we examine a few of these, it is important to look again briefly at the question of the logical foundation of the thesis of neutrality. The conventional argument, dating from Hume (see earlier in this chapter) claims that science is concerned exclusively with what *is*; it operates only with positive statements, never with normative statements which concern what *ought to be*. According to this view, a normative conclusion cannot be derived deductively from positive premises because it is not already implied in those premises (see chapter 2); therefore science cannot give rise to normative statements (value judgements).

This orthodox position asserts that while science can present the factual evidence of any connection, let us say, between smoking and the health of a mother and her unborn child, it has nothing to say on whether it is right for the individual mother to smoke. Despite the statistical evidence that prolonged heavy smoking reduces the life expectancy of the smoker as well as the survival prospects of the baby (and despite the performance of controversial experiments on 'smoking' animals, the *justification* for which is supposedly the alleviation of human suffering and the saving of human life), science itself simply cannot pronounce on whether the individual *ought* to shorten her own life or adversely affect that of her baby. Equally on this view, science as such can offer no guidance on whether a nation ought to engage in genocide (say through the agency of nuclear war). 'As between favouring genocide and opposing genocide, science is neutral; it has, one might say, no moral dimensions' (Black).

In discussing such questions, the philosopher Max Black wishes to attack the classical position. The pillar for his argument is the claim that it is not, after all, always possible to distinguish infallibly between positive and normative statements. For example, the proposition that murder is a sin certainly implies that one ought not to murder, yet one cannot tell from the linguistic form alone whether the proposition is normative or positive. Black argues that some categorical normative statements 'do have truth-value, and can be certified as true objectively (independently of desires or hopes)'. In support of this he invokes the anthropological evidence that all human beings, regardless of background, appear to share certain basic ethical principles, for example the principle that it would be wrong to eat children when food is scarce. 'Even if factual truths had to be regarded as logically segregated, the introduction of some generally acceptable normative (premises) would legitimate the derivation of normative conclusions.' Black's own overall conclusion is that science 'as a system of voluntary activities . . . is, at least in principle, amenable to internal or external control'.

It is obvious that science can be, and is, controlled from outside, for although it can devise efficient means for committing genocide, whether or not genocide is committed is largely a matter for society, not science. But the suggestion that science may, after all, have an *internal* ethical dimension, is one of great interest and one which forms a good basis from which to look at some of the pressing issues of the present day. The literature on these subjects is already enormous and is expanding rapidly. For present purposes we will begin by identifying a few of the general ecological problems relating to science and technology and then go on to examine some specific examples involving the physical, life and social sciences. It is as well to note that the moral questions at stake are sometimes only implicit in issues which are discussed more at the level of those conditions which may be 'caused' or 'cured' by science.

'Spaceship Earth': Science, Technology and the Future

The most alarming prospect faced by a world which appears in many ways to be increasingly unstable, must surely be the obliteration of the human race. This has by now been said so many times that it sounds almost like a cliché, yet the fact remains that the modern science-related military-industrial complex has, through its discoveries and inventions, instigated a qualitative change in international relations of the most far-reaching kind. The continuing acceleration of the nuclear arms race is officially justified on each side by the need to maintain a 'balance' which is supposedly threatened by the other, yet it is considered by much informed opinion to have increased the hazards of global war, without reducing its likelihood. (Aspects of *Science and War* are discussed at the end of this chapter.)

Of less immediate threat, but of immense seriousness in the relatively short term, are the instabilities in the material basis of society that appear to result from the impact of science. Advances in the standards of medicine and public hygiene have permitted a dramatic growth of world population by reducing disease and child mortality. Meanwhile, agricultural science has increased food production to the extent that this huge population has been fed – albeit at varying dietary standards and with some horrifying, and it seems increasingly frequent, exceptions. In the richer parts of the world, technological society has increased people's expectations of a high standard of living, and through disproportionate allocation and consumption of resources, led to strained relations between the developed and under-developed countries (the North-South dichotomy). In particular, it is argued, science and technology increasingly concentrate on the production of luxury goods for the affluent few, and on the military equipment to protect this affluence, while ignoring such basic needs as food, shelter, clothing, health, education and employment for the powerless majority. Furthermore, by imposing the technological values of the affluent world, fundamental democratic principles such as popular participation and control, are being undermined in the poor countries of Africa, Asia and Latin America as the requisite sophisticated expertise is concentrated into fewer and fewer hands.

Many modern science-related industries produce new threats of pollution which endanger the delicate ecological balance of nature, perhaps irreversibly and irremediably. For example, DDT destroys disease-carrying and crop-destroying insects but, since it is non-selective, it kills many other forms of life as well. Furthermore, it is persistent in nature and has spread, through air and water, to all parts of the earth. It accumulates through food-chains and is already known to effect fecundity in birds, although it has not yet been shown to be harmful to man at the present levels that have accumulated in his tissues. There are many other examples of global importance, such as the danger of increased harmful radiation from the sun as a result of our depleting the ozone layer of the upper atmosphere through the use of flurocarbons as propellants in aerosol cannisters, or the precipitation of 'acid rain' which can upset the balance of lakes and rivers and is thought to result from the output of sulphur into the atmosphere, typically from coal-fired power stations. Many such cases are notably ambiguous in the ethical context, because those who suffer the consequences of pollution are frequently not those who produce it.

The problem of pollution is closely related to the phenomenon of technological growth and so to that of the increasingly rapid exhaustion of natural resources, especially rare minerals and fossil fuels. It has been estimated that the total consumption of minerals was greater in the first fifty years of the present century than during the whole of preceding time and that between 1950 and 1975 even this rate was surpassed by a further 50 per cent. Consumption of petroleum has risen by a factor of at least 100 during the twentieth century. These figures give rise to very real fears, not only of an 'energy crisis', but of a crisis for our very industrial life-style also. This raises many questions as to what sort of society we ought to seek to establish, as well as those concerning the society we may be obliged to accept. Since all these problems seem to be bearing down upon us at an increasing rate, many of the present ways of coping are likely to be inadequate in the future. As yet, opinions differ as to whether future technology will be able to resolve the problems caused by present technology (the so-called 'technological fix') or whether we are, perforce, entering a 'post-manufacturing' age in which such qualities as asceticism and austerity will be valued more highly than productivity and economic growth. According to the latter view, we should begin consciously to adjust to a new society which will emphasize the 'quality of life' rather than the quantity of production.

Either way, there is no disagreement that grave problems lie not far ahead. This realization has promoted new ways of thinking about the future, and especially about the next 20 to 100 years. The established social sciences, such as economics, have generally avoided making projections so far ahead, but it is in this period that problems now being confronted would be expected to become most acute. As a result, a new social science of futurology has emerged, building upon techniques of technological forecasting, Unfortunately, while trained natural scientists and technologists have played a prominent part, futurology has revealed many of the classical difficulties encountered in the development of any effective social science (see the end of chapter 5). For example, the futurologists can deal only with *possible* futures, not with what will inevitably happen; that is, they cannot actually *predict* the future because they can never anticipate revolutionary discoveries or inventions, complex interactions between social factors, or the ways in which people may react to their predictions, perhaps falsifying them. They are working on such a protracted time scale, that they can learn only very slowly from their successes and failures. In the meantime, however, they ask that we act *now* to forestall the dangers they foresee.

The attempts to construct 'world models' which simulate the total situation on computers, have turned out to give results that are highly sensitive to the values of factors which can only be guessed. The most famous example was presented in *The Limits to Growth* (1972), produced at the prestigious Massachusetts Institute of Technology by Dennis Meadows and his colleagues. Essentially, the book is an attempt to combine the effects of the population explosion, the pollution of the biosphere and the depletion of natural resources, by a sophisticated mathematical model which, supposedly, then allows the forecast of

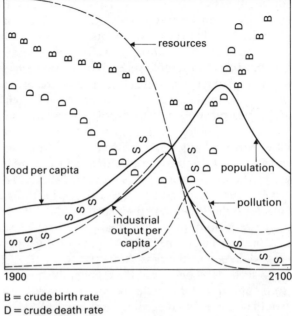

B = crude birth rate
D = crude death rate
S = services *per capita*

Figure 9   *The 'standard' world model*

future trends from this integrated basis. In the 'standard' world model (figure 9) it was assumed that the traditional relationships which have operated in the world will continue to operate in the future. On this basis, it was forecast that the industrial foundations of society would collapse early in the twenty-first century owing to the increasing price and decreasing availability of natural resources. Food per capita would then rapidly decline, while pollution levels would continue to rise for some time. Finally, there would be a precipitous fall in world population through famine and disease. The only way to achieve a stable, as distinct from an unstable, model would be, according to Meadows, to assume 'unlimited' resources (through the exploitation of low grade ores using cheap nuclear power, and extensive recycling), effective pollution controls and a huge, though feasible, improvement in food production. Then if birth control also limited the average family size to two children, and industrial output per capita were limited to 1975 levels, collapse could be avoided in the long term (figure 10). A delay in implementing the 'necessary' changes from 1975 to the year 2000 would, predicts the model, again result in instability and collapse by the end of the twenty-first century.

Meadows' study has been criticized severely by subsequent authors on

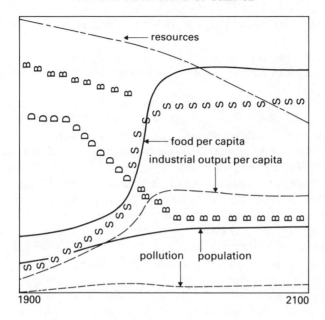

Figure 10 *A stabilized world model*

a number of grounds, principally that a 'world model' is impossibly general; that it ignored the effects of economic forces and technological advance; that it underestimated the availability of resources; and that it was generally too pessimistic. But whatever shortcomings the study may have, both it and the responses of its critics illustrate the enormous difficulties inherent in any attempt to predict future events.

Physical Sciences: Nuclear Power

The vigorous debate over nuclear power, which has always been linked to the debate on nuclear armaments, is perhaps the prime example in modern times of the agonizing problems raised by science and technology. The controversy has been lent a new urgency by the depletion of conventional fossil fuels. A prompt decision to press ahead with a full-scale nuclear reactor programme – especially if it involves the development of the fast breeder reactor – represents, according to the enthusiasts, the *only* viable solution to the overall energy crisis, a solution which would be the magnificent fruit of scientific and technological progress. To the critics on the other hand, such a decision would be a disastrous and, at least for the foreseeable future, an unnecessary, example of society's subservience to the 'imperatives of technology'; disastrous because the safety and efficiency of the technology is still unproven, and unnecessary because the use of safe, renewable energy sources, coupled with efficient conservation, could adequately supply future needs. Although the controversy is sometimes debated on the technical and economic level only, almost all of the issues have underlying ethical dimensions of enormous social significance.

Uranium is not a fossil fuel in the usual sense, but neither is it inexhaustible. Since it is the fuel of thermal nuclear reactors (see the previous chapter) this is therefore a problem. It is recognition of this that prompts some experts to advocate the fast breeder reactor, a type which is fuelled by plutonium – produced in conventional thermal reactors – and also 'breeds' more of it for use in other reactors. (The first full-scale reactor to use the breeder principle was that at Dounreay, in the north of Scotland.) Unfortunately, the whole question of the availability of suitable uranium is itself highly controversial. Some 'authoritative' estimates claim that reserves are sufficient to allow indefinite postponement of a breeder programme, while others doubt that they are adequate to support a large breeder programme at all. Obviously, as lower and lower grade ores have to be mined, there must come a point when more energy is consumed in getting the uranium than can be obtained from using it as a fuel. Quite apart from the environmental issues raised by widespread mining, one ethical dilemma behind this is the radiation sickness and death suffered by those who mine the ores. These men have little influence and often little choice of occupation, and their exploitation is, according to the critics, 'one of the great scandals of our time'. As for plutonium, not only is it the man-made element essential for nuclear weapons – hence the fears of proliferation – but because of its radioactivity it is also one of the deadliest substances known. A large breeder programme would necessitate the frequent transportation of plutonium, with the inherent risk of accidents and terrorist attacks. A vast body of armed police would therefore be essential to guard the trains and the reactor sites, a prospect which gives rise to fears of the incompatibility of nuclear power and individual liberty.

The question of safety is the one of most concern to the anti-nuclear movement. There are several aspects of this problem, including the possibility of catastrophic accidents within nuclear power reactors, and the long-term effects of an increase in the 'background' radiation which is the feared result of leakage from operating plants and waste products. The recent prospect of a major disaster at the 3-Mile Island reactor in Pennsylvania was not in the event fully realized, since the overheating of the core did not result in a 'melt-down' of the fuel elements sufficient to form an explosive pool of liquid fuel at the base of the reactor. In the two major meltdown accidents which have occurred to date in the West, the reactors' containments were not broken, and the proponents of nuclear power argue that, with appropriate design, such breakage is virtually impossible. However, the 1986 disaster at Chernobyl in the Ukraine did little to reassure the public that such design is yet universal, something it clearly needs to be. An uncontrolled meltdown accident could be a genetic 'timebomb' with hideous consequences over many generations, and it could cause the deaths of many thousands as a result of the escape of radioactive material. It would certainly require the evacuation of huge areas of land for centuries. According to controversial reports there *was* a major accident at a nuclear waste dump in the USSR in 1958 which has resulted in the removal of some thirty communities from the map. It is alleged by critics that this was concealed from the public in the West for over twenty years by means of a conspiracy of silence on the part of the nuclear industry.

The disposal of radioactive waste also poses great problems. Leakage from Britain's plant at Sellafield has been described as unstoppable, and the Irish Sea is now claimed by some to be the most radioactive in the world. Some waste products will remain lethal for tens of thousands of years and the technology for their disposal has yet to be developed. At present the plan is to fuse the waste materials in a special type of glass and to bury the containers in geological formations of proven stability. But again, this gives rise to grave anxieties and to the question of our moral responsibility to leave a healthy environment for those who come after us.

Of course it is because nuclear power stations are potentially so dangerous that they cannot be sited near to areas of high population. This means that the vast quantities of waste heat produced cannot be applied usefully, say for community heating. Electricity, which the stations produce, is widely agreed to be highly uneconomic for space heating, yet on the UK Department of Energy's own estimate, we can expect more than 50 per cent of the nation's primary energy to be dissipated as waste heat from power stations by the year 2020. According to the critics of nuclear power, the enormous expense of research in nuclear technology, means that little finance is available for that into alternative sources of energy (such as the sun, wind, tar sands and so on) and into conservation; for example, it has been calculated that about twice as much energy can be saved by heat conservation as can be produced for the same financial outlay.

There is already a very extensive literature on these great problems and we have touched on only some of them. Perhaps the single most important point that should be made is that the opinions of the scientific experts differ because in the last analysis they simply cannot be made value-free; it is impossible to weigh the benefits and hazards of nuclear power within the narrow confines of quantifiable parameters alone. Thus, although a distinction is sometimes drawn between 'problems in ecology', which are supposed to be the concern of scientists alone, and 'ecological problems' which are chiefly political and social, there is seldom in practice a full separation. For instance, the scientist can measure the level of radioactivity in the air or the sea and obtain the unqualified agreement of all other scientists on the question of the accuracy of his measurements. But problems soon arise when the scientist is asked to advise the politicians on what level of radioactivity is 'permissible' or 'acceptable'. The answer to such a question draws the scientist into conflicts over values, which can only mean into areas beyond his special authority as an expert. As a scientist, it may be argued that he is obliged to confine himself exclusively to the measurements, yet it is also as a scientist that he alone may be able to make considered recommendations. As a citizen, he may feel a moral obligation to provide advice against a realistic background provided by the social and political issues involved, and these are bound to reflect his personal views.

For the economist the problems are likely to be even more difficult. His basic data are almost certain to be more controversial than those of the natural scientist and in his wish to produce accurate forecasts he may be tempted to extend his economic analysis to human values which, for many people, cannot be quantified without losing their essential humane character. How, for instance, should the economist attempt to calculate the real cost of nuclear-powered electricity? Should he take into account the vast expenditure on research, and on reactor failures, and should he try to estimate the cost of 'retiring' and sealing reactors at the end of their productive lives? Even more controversial, should he try to measure the health and environmental factors in purely economic terms, or should he leave the importance of these to be judged by the politicians? In short, should his brief be restricted to what is likely to be efficient for society or is he, like the politician, obliged also to consider what may be *good* for society as well?

Life Sciences: Animal Experiments

In recent years there has been a vigorous renewal of the debate concerning the use of animals in the service of man. Much of this has centred on the question of whether experimentation on animals 'for the benefit of mankind' can indeed be justified. For those who take the Cartesian view that animals are intrinsically of little worth, perhaps coupled with the Kantian idea that only the human individual exists as 'an end in himself', there will be little problem. Experiments, even painful experiments, will certainly be justifiable if there is obvious human gain, and they will probably also be justifiable if the purpose of the experiment is no more than the 'advancement of knowledge'. However, this traditional philosophical position is under threat from some quarters, many people being no longer willing to draw a clear dividing line between man and the rest of nature, in particular between man and his fellow animals. To such people, it seems that many of the experiments now performed are ill-conceived and unnecessary. Under the influence of writers such as Ivan Illich (in his book *Medical Nemesis*), they are inclined to question the basic assumption that large-scale improvements in human health can in any case be related at all clearly to advances in the paramedical sciences. Rather, they seem often to be the result of elementary improvements in public hygiene.

The medical establishment, however, still exercises enormous influence. Orthodox opinion continues to be that, while suffering should be minimized, it is nevertheless essential that many millions of experiments be performed on living animals each year. A proportion of these are for purposes quite directly related to medicine and surgery, but a very large number merely 'test' the vast range of new products, from shampoo to floor polish, which are continually being made available to the public. One experiment of the latter kind, which is very widely practised and legally required, has been the subject of intense controversy. This is the so-called LD_{50} toxicity test. Any material which might be poisonous has to be force fed to animals, usually rodents, rabbits and dogs, which are then left to die in order that the 'lethal dose' – that which killed 50 per cent of the animals – may be determined, by extrapolation, for man.

The subject of experiments on animals is too large for us to examine in detail. One thing, however, is certain. This is that it is not just a matter of emotion and sentiment, but raises ethical questions of the profoundest kind. For our main purpose – that is, for consideration by scientists themselves – it is perhaps most useful if we briefly explore the case critical of orthodoxy, because its fundamental claims generally receive little hearing among those 'indoctrinated' by the prevailing wisdom.

A view commonly expressed by those who practise animal experimentation is that of D. H. Smyth in his *Alternatives to Animal Experiments* (1978). This book presents a reasoned account as to why the *status quo* should continue more or less unchanged. The moral issues, however, are omitted, says Professor Smyth, 'not because I have no interest in these or consider them unimportant, but because I have no special qualification to instruct others on these topics.' This is fairly typical of the scientist's position. Quite apart from the argument that an individual unfamiliar with the basic moral issues is not thereby absolved of responsibility, there is the danger that this attitude may quickly become one which implies that because there is no logical connection between 'objective' matters of fact and subjective matters of opinion (see above), the moral issues themselves do not really exist, or at least that if ignored they will go away.

Perhaps the most important moral question to be asked here is: Do animals have rights (let us say, to life, liberty and the pursuit of happiness) in the sense that human beings are generally thought to have? The answer to this may be in terms of the answer to another question which is more based in empirical evidence: Do animals have interests? If we restrict ourselves to mammals and birds, which represent the great bulk of animals used by para-medical scientists, there would seem to be no particular problem. For example, if we believe that other human beings feel pain as we ourselves do, there is no physiological or philosophical reason to suppose that mammals and birds do not also feel pain in much the same way. They have, we may therefore say, an interest in not being made to suffer pain. However, that a creature has interests does not logically imply that it also has rights, but only that it *can* have rights. Whether or not it is thought that the creature does indeed have rights will depend upon our perception of the *nature* of that creature. This is why the most fundamental kinds of rights are often described as 'natural' rights, and the particular natural rights of human beings are termed 'human rights'. Our concern, in this same sense, is with 'animal rights' and with the arguments that may be offered as a basis for any distinction between them and human rights. The question of animal rights (in particular the right not to be caused unnecessary suffering) is crucial to the whole question, because if we feel obliged to recognize such rights we are obliged also – if we are concerned to be morally consistent – to advocate drastic changes in our treatment of animals. Incidentally, even those individuals who have reasons to reject the concept of animal rights, do not thereby have licence to be cruel to animals, for there would still be the question of our duties to beings without rights.

The essential element in the argument for the rights of animals is that there is no condition possessed by *all* human animals and which is an agreed basis for their having rights, which is not also possessed by some non-human animals, save the condition of 'being human' itself. Conversely, there is no condition the lack of which in *all* non-human animals might serve as a basis for their not possessing rights, which is not also lacked by some human beings. For example, the traditional capacities for language and reason, used since ancient times as a basic distinction between human and animal nature, are no longer considered by ethologists and animal psychologists as the unique monopoly of man, and in any case they are not possessed by human infants, nor even potentially, by some human imbeciles. Even if the possession of such capacities were denied to non-human species, this would not be a logically valid reason for denying them fundamental rights such as those to life and liberty, although it would of course be a sufficient reason for denying them the right to vote. Thus there is no inconsistency in affirming that there *are* rights which we human beings have, not simply by virtue of being members of the species *Homo sapiens,* but rather as a consequence of certain capacities or interests possessed uniquely by human beings. But to make the mere difference of species *the* basis for a defence of grossly different treatment is, it is claimed, precisely analogous to the argument that justified slavery or Hitler's 'final solution' on the basis of differences in race, or polygamy and serfdom on the basis of differences of sex or caste. What seems assured, according to this argument, is that the concept of 'human rights' is a great deal more restricted than is conventionally believed; because non-human animals have an interest in experiencing pleasure and in avoiding pain, they have natural rights to life, liberty and the pursuit of their particular happiness in much the same way as human beings do.

To those who are critical of this view and who continue to insist that there is some uniquely intangible essence to human nature which transcends man – the extraordinarily highly evolved mammal – and which might therefore be a moral basis for according different treatments, the answer offered by its advocates is that it is this very essence which makes of man, uniquely among all species, the 'ethical animal'. It is surely the case, they assert, that man is alone in bearing the burden of 'moral agency', so it is he alone who is outside of and above the unthinking 'law of the jungle' which sanctions the dominion of the strong over the weak. According to this view, the essential essence of man's humanity is not a biological conception at all, but rather a spiritual one which he alone can contemplate and through which he alone can identify the kind of behaviour to other creatures that is uniquely true to himself.

In summary then, a view such as that outlined above – still that of a tiny minority – argues that *if* human beings have interests and rights, so too do animals. It particularly asks of the orthodox scientist who is prepared to exploit animals how, in the light of evolutionary theory, there is a special intrinsic dignity attached to his own species but apparently not to others. If he argues that there is some more tangible characteristic of man, he must be sure that this refers *only* to human beings and not also to closely related species, otherwise he may find it difficult to justify experimenting on apes (and, for that matter, eating pigs), while simultaneously condemning as outrageous the suggestion that we might do likewise with human babies or imbeciles. At any rate, some would say, what in good faith he cannot do is let gross inconsistencies in his own moral position go unchallenged or unresolved.

This unorthodox viewpoint cannot be ignored. No self-respecting scientist will be content to dismiss it as eccentric or 'cranky', for this will merely suggest his inability to refute an argument which poses a clear threat to the conventional position and demands a thorough analysis of the inner ethical dimension of physiological science (see above).

Social Sciences: Deception and Degradation

The various codes of practice followed by biological scientists who work with animals typically gloss over the more fundamental ethical dilemmas of the kind discussed above. For scientists working with human beings, namely, medical and social scientists, recommendations of a rather more explicit sort are usually available, an example being those of the National Commission for the Protection of Human Subjects in Biomedical and Behavioural Research in the United States. These, however, are regarded by most social scientists as having been designed primarily for medical research, and as being either too inflexible or wholly inapplicable to the complex array of situations facing, in particular, sociologists and psychologists.

Scientists working in these disciplines therefore tend to have their own ethical principles. We shall examine two 'classic' studies of the 1960s in which these social science codes of conduct were put under severe strain. Each of them has been the subject of immense controversy, the essential question being whether or not it is justifiable to deceive, and perhaps to degrade, a relatively small number of human individuals in order to gain information and understanding on issues of widespread public importance. Is it, we might ask, justifiable to use human beings as 'means' to some end, or is even the disinterested scientist obliged, in conformity with the Kantian imperative, to treat them always and only as 'ends in themselves'?

The first investigation concerns the ethics of 'the sociologist as voyeur'. In the book *Tearoom Trade* (1970), Laud Humphreys describes a study he undertook into the sociology of homosexual behaviour by observing impersonal contacts in a park restroom or public toilet (the so-called 'tearoom'). Humphreys started out in the way common among students of 'deviant' behaviour, by 'going into the field' of his interest. However, he soon began to concentrate his attention on the 'tearoom' because it 'could make a useful contribution . . . [and] seemed to be a unique social setting . . . worthy of study in itself.' In particular, it provided a 'means for direct observation of the dynamics of sexual encounters *in situ*, . . . [and for] the gathering of a representative sample of secret deviants, for most of whom association with the deviant subculture is minimal.'

Humphreys chose his methods 'because they promised the greatest accuracy in terms of faithfulness to people and actions as they live and happen. These are strategies that I judged to be the least obtrusive measures available – the least likely to disturb the real world.' First he posed as a 'watchqueen' or look-out for the homosexuals, guarding them from potential intruders for short periods, but secretly making copious notes on their behaviour with the aid of a tape recorder concealed in his car parked outside. After some time he was able to conduct confidential interviews outside the tearoom with selected subjects in order to fill out his own observations. But a rich source of social data was obtained by surreptitiously noting the licence numbers of his subjects' cars and representing himself as a 'market researcher' in order to gain access to the licence register held by the police. Having thus collected names and addresses, he spent a Christmas vacation on the streets, noting all relevant details of home, neighbourhood, marital and family status and so on, in order to compile a more comprehensive dossier on each subject. Finally, and some time later, it happened that Humphreys was asked to develop a questionnaire for a social health survey, and he took this opportunity to add his 'deviants' to the over-all sample for the survey. He was thus able to compile yet more information on family background, health, religion, employment, sexual attitudes, and so on, while interviewing them – his own appearance having been carefully changed – as representing 'normal' people.

Throughout his study, Humphreys took scrupulous care to preserve the confidentiality of his subjects, and his book represented no threat to homosexuals, portraying them as ordinary individuals who themselves pose no threat to society. The investigation received a major award from the Society for the Study of Social Problems, yet its methods obviously involved sustained lying and deception and took grave risks with the self-respect and reputations of its subjects. Was it justified? Humphreys himself asked the question : 'Are there, perhaps, some areas of human behaviour that are not fit for social scientific study at all? Should sex, religion, suicide, or other socially sensitive concerns be omitted from the catalogue of possible fields of sociological research?' For his part, Humphreys certainly felt that his end – not only the growth of knowledge, but the improvement of the position of the homosexual in America – was ample justification of the means he employed. By no means all social scientists have agreed with him.

The second classic investigation is that described in *Obedience to*

Authority (1974) by Stanley Milgram. The purpose of this psychological experiment was 'to find when and how people would defy authority in the face of a clear moral imperative.' Its social importance was judged by Milgram in terms of a contribution towards the understanding, and hence the avoidance, of a situation like that which prevailed in the Nazi's attempt to exterminate the Jews. But to what degree a simplified experimental situation can genuinely claim to duplicate a real-life one is, of course, a matter for debate.

The experiment involved three people: a 'teacher', a 'learner', and the experimenter himself (the authority-figure). The teachers were all recruited on the basis of an advertisement in a local newspaper (and paid for their services). One teacher and one learner performed in each session and lots were drawn to decide who should play each role. However, unknown to the naive subject, the method of drawing lots was rigged to ensure that he or she was always the teacher, while the learner (or 'victim') was, in fact, an accomplice, carefully trained to act his role. The experiment concerned the effects of punishment on learning. The role of the teacher was to give electric shocks of increasing severity to the learner each time he made an error in a simple learning task. In reality, the learner received no shocks at all, but he did provide graded, audible 'feedback' to the teacher in the form of mild grunts to apparently agonized screams, in proportion to the strength of the 'shock'. Before commencing the session the teacher was himself given a mild shock to enhance his own belief in the authenticity of the shock-generator over which he had control, and he also witnessed the learner being strapped into an 'electric chair'. Once under way, however, the teacher could no longer see the learner, though he was in the presence of the stern, impassive experimenter. The latter informed the teacher that he was required to present the learning task to the learner in the next room and, for each wrong answer, administer an electric shock, beginning at 15 volts (slight shock) and increasing for each subsequent error in 15 volt intervals, through 'strong shock' and 'intense shock', to 'danger: severe shock'. The point of the experiment, said Milgram, was thus 'to see how far a person will proceed in a concrete and measureable situation in which he is ordered to inflict increasing pain on a protesting victim. At what point will the subject (teacher) refuse to obey the experimenter?'

The experiment gave rise to intense and obvious conflict on the part of the teachers.

> On the one hand, the manifest suffering of the learner presses him to quit. On the other, the experimenter, a legitimate authority to whom the subject feels some commitment, enjoins him to continue. To extract

himself from the situation the subject must make a clear break with authority.

The one concrete measure made on the teacher was the maximum shock that he was prepared to administer before refusing to co-operate further. At the end of the session, he was reassured that the learner had in fact received no shocks and, no matter whether he had been defiant or obedient in the presence of the experimenter, that his behaviour was perfectly normal. Thus, some attempt was made to avoid dangerous guilt on the part of the teacher on learning perhaps unsavoury aspects of his own nature, yet it is towards this aspect of the research that much censure has recently been directed (as well, of course, as to the matter of deception itself). The prime moral question here is therefore that of whether the experiments involved unjustifiable degradation of the subjects as they confronted unsought self-knowledge.

Milgram's investigation, like that of Humphreys, was at first received with acclaim, although in recent years it has been widely condemned. This shift of opinion reflects a fairly general move away from the view that the rightness or wrongness of an action can be assessed in terms of its consequences (such that the discomforture of the subjects in Milgram's experiments was justified because of the contribution it made to our understanding of human nature), towards the view that some actions are intrinsically wrong, no matter what the potential benefits may be. The controversy is one which social scientists in particular are increasingly obliged to confront.

MORAL ISSUES IN SCIENCE AND WAR

As a final section in this chapter it is important that we examine briefly some of the ethical questions which arise in the relations between science and warfare. During the 1960s and 1970s a good many of the 'anti-science' feelings which were generated (see the next chapter) undoubtedly stemmed from the widespread association of certain scientific activities with war and the preparation for war. That there is such an association is undeniable (see the previous chapter); that it should give rise to distaste and disillusionment is hardly difficult to understand.

The association is not in itself a modern phenomenon, although its scale since the Second World War is something altogether new. Since ancient times the skills of technically-minded men have always been harnessed to the needs of their nation in times of danger (we might think, for example, of the mirrors and seige machines of Archimedes) and it is not an exaggeration to say that the origins of science in the modern sense have a good deal to do with war. In the engineering of Leonardo, the hydrostatics of Stevin, and the mathematical mechanics of Galileo, we see the emergence of the scientific and technical 'expert'. From the late nineteenth century – that period of prolific technological innovation which saw the development of the railway, the steamship, the telegraph, explosives and so on – we see the beginnings of the modern arms race. Whatever detailed ethical questions may have been raised at that time, the essential guide to justify action was, and indeed still is, the principle: 'If we don't do it, the other side will.'

We are concerned here not to catalogue the innumerable cases of the influence of science on war, but only to identify those moral elements in their reciprocal interactions which cast so dark a shadow over the modern world. Despite a universal fear and horror of war, these interactions continue to accelerate, apparently because they are 'inevitable' in a world that is dominated by science-related technology and which so manifestly lacks a 'covenantal ethic' of mutual trust and obligation (see pp. 146–147).

A famous expression of the view that close relations between science and the war machine are not only inevitable, but actually desirable, is the conclusion of the United States Government Study Group, which in 1968 published its findings under the title: *Report from Iron Mountain on the Possibility and Desirability of Peace* (Lewis).

> War is the principal motivational force for the development of science at every level, from the abstractly conceptual to the narrowly technological. Modern society places a high value on pure science, but it is historically inescapable that all the significant discoveries that have been made about the natural world have been inspired by the real or imaginary military necessities of their epochs.

This is obviously an extreme view which can certainly be refuted in specific instances. Nevertheless, it is also by no means without support. Certainly since the First World War, numerous scientific developments can be traced back to origins in so-called defence research. For instance, the mobilization of scientists which eventually occurred at that time led rapidly to the application of fundamental principles of physics and chemistry to work on such devices as piezo-electric microphones for submarine detection, and to the development of poisonous gas as a battlefield weapon.

Gas warfare was the most famous as well as the most controversial development of the period. Although the Germans – who at the beginning of the twentieth century were overwhelmingly powerful in engineering and technology – were the first to use gas, the British and French soon retaliated in kind and with little concern for any ethical issues that may have been involved. Chlorine, phosgene and mustard gas (dichlorodiethyl sulphide) were used on a large scale, causing over one million casualties, but of particular interest in the present context was the claim by many scientists that gas warfare was actually more 'humane' than that conducted with explosives. Although gas could be crippling over a wide area, permanent disabilities were often less serious, and full recovery more common, than with conventional weapons. This was not the official view, however, chemical warfare having been outlawed by the Hague Convention as early as 1899, and again by the Geneva Protocol of 1925. In the 1930s, the development of 'nerve gases' of almost incredible potency introduced an altogether new dimension, but thus far the horrifying potential of such weapons appears to have operated as a mutual deterrent, indicating that at this ultimate level (as also with strategic thermo-nuclear weapons) the fear of war can be an 'effective' fear which, for purely realistic reasons, produces the restraint which might not have come from ethical motives alone.

By the time of the Second World War it was clear to all that the outcome of the military conflict would be decided by the applications of science. The likelihood of bombardment from the air proved to be the essential stimulus that led to the invention of radar. This device gave to the Royal Air Force aerial superiority in the critical Battle of Britain, and ever since has been the backbone of all the systems designed to provide early-warning of attack by enemy aircraft or missiles. The idea of using pulsed radio waves, which could be detected after reflection back from an aircraft, occurred more or less simultaneously to physicists in Germany, Britain and the United States, but it was the invention of the cavity magnetron by a wartime team of 'pure' scientists working at Birmingham University which made possible a really sophisticated system. The magnetron could generate a high-powered beam of radiation at short wavelengths (microwaves) and its development by the Allies gave them a critical advantage in aircraft detection and interception. Once again there were few, if any, voices of dissent among the scientists, for the terror of the Blitz and the unthinkable horror of Nazi victory quickly dispensed with any doubts as to the moral justification of resistance and retaliation.

The concept of mobilization in the war effort, not just of individual scientists but of science itself, reflected a wholly new attitude of government to science, one that has been maintained ever since. In the United States this change was especially dramatic. Whereas before 1940 there had been 'mutual aloofness' between science and government, the effect of the war itself was to bring home to the scientists in particular the realization that no matter how ambivalent their own feelings about whether science needed government, there was no longer any doubt that government urgently needed science. As Daniel Greenberg puts it:

> The unsuccessful supplicant was embarked on saving the rejecting patron, and it can be reasonably argued that the long-term orphan did precisely that, with a display of persuasiveness, political skill and technical performance that comprises one of the most arresting chapters in the nation's history . . . .

That the orphan blossomed into the very acme of political and military power was hastened by the catastrophic incident at Pearl Harbour. This led to the establishment of the Manhattan Project in June 1942, a huge and intricately organized effort run by the US Army, to design and build an atomic bomb. The extraordinary rapidity with which it achieved its goal represents more than just *the* classic example of the effect of the pressures of war on science (for there was a desperate fear that the Germans might produce the bomb first); it instigated a totally new situation in international relations for, with the subsequent development of the hydrogen bomb, man had within his power the capability to annihilate the whole of civilized life on earth. It is hardly surprising that this raised for the scientists themselves, ethical dilemmas of the profoundest kind.

It was not merely that the scientific imperative of universalism or internationalism was no longer compatible with the war-borne dependence of science on the state. Rather, it was that for the first time in history science had become *the* agent of the state and for the individual scientist there was the agonizing problem of two-fold loyalty, on the one hand to the co-operative spirit of science, and on the other to the competitive needs of nationalism. Any remaining vestige of moral or political disinterestedness had been swept away by the sensational success of the Manhattan Project, and, as we have already said, the initial problem of whether the atom bomb *should* be built was never seriously asked in view of the advanced state of the hostilities and dire suspicions concerning the research activities of the 'other side'. As to whether the bomb, having once been built, should actually be used against the Japanese, there was in fact a great deal of soul-searching in the United States, and sincere differences of opinion. In the event however, no clear warning was given about the bomb's potential, the idea of a harmless demonstration having been abandoned as impracticable because it was not certain that the weapon would actually work. The destruction of Hiroshima, and then of Nagasaki, was carried out for the military purpose of halting the Japanese war as soon as possible.

American casualties were thereby minimized and, according to most experts, the Japanese casualties (more than a quarter of a million dead) were probably fewer than might have been expected as a result of conventional air raids and invasion. But the horror of the devastation, and the long-term teratogenic consequences, left ethical questions which remain unanswered to this day.

The broad support for the government shown by the American scientists in 1945 was beginning to falter by the early 1950s. The new dimension was the hydrogen bomb. Whereas with the atomic bomb – which is the product of the fission of atoms of heavy elements – there are theoretical limits to size and therefore to destructive power, with the hydrogen bomb – which depends upon the fusion of atoms of light elements – there is no such limit and the destructive potential is truly genocidal. By this time, the world had entered a period of 'cold' rather than actual war, and without the alarming urgency wrought by the latter, the question of whether the bomb should be developed at all was on this occasion extensively debated. The scientific advisory panel of the US Atomic Energy Commission, chaired by J. Robert Oppenheimer – the war-time director of the Manhattan Project's main laboratory – came out against development, predominantly on the ground of the intolerable risks it presented to the survival of mankind. The panel was thus motivated by universal rather than narrowly nationalistic concerns, but it is only fair to add that they were acting from a position of considerable strength. It was believed that American superiority in nuclear weapons technology was then so marked that an act of renunciation, which might conceivably have forestalled the spread of nuclear armaments, would in any case have posed no threat to national security. However, their position was over-ruled by government; the hydrogen bomb was built and tested; and Oppenheimer himself suffered the tragic indignity of dismissal from his post and cancellation of his security clearance.

Oppenheimer's painful involvement in the moral and political debate instigated a new era in which scientific experts not only came to accept their new 'politicized' role, but actively to espouse their personal views. In this respect it is important to note that while Oppenheimer was undoubtedly cast as hero and martyr by the liberals, he received by no means universal support from his fellow-physicists; indeed, one of great influence who opposed him was the so-called father of the hydrogen bomb himself, Edward Teller. Thus there were deep and damaging divisions within the ranks of the scientists in much the same way as between scientists and politicians, but the remarkable feature of the new era was the overt dissent of some of the outstanding scientific authorities. For them, the spirit of co-operation fostered by science was a matter of personal experience; a sense of international community transcended the fear and aggrandizement which they saw as the springs of competitive nationalism.

The dissenting movement was to find expression in the Pugwash association (which first convened in 1957 in the Nova Scotia village of that name) at whose annual meetings there are detailed discussions between Eastern and Western experts on matters concerning disarmament and world peace. Pugwash has had quiet but significant influence on the political relations of the Great Powers, while more specific movements have concentrated on such issues as anti-ballistic missile defence systems, biological and chemical warfare, and the status of space in connection with the possibility of war on earth. To end on an optimistic note, it begins to appear that a new 'consciousness' is emerging and, as John Ziman puts it, 'Big Science has grown to maturity in the United States, and is now learning the burdens and moral responsibilities of being an Estate of the Nation.'

8

Science, Culture and Religion

No-one today would seriously argue that science can operate in an intellectual vacuum, guided only by its internal logic and universal ethos. Indeed, we saw in earlier chapters that there are many examples of where the idealized model of science, as a 'pure' and self-regulating system which exchanges knowledge for recognition, has broken down, simply because its independence of other forces cannot be maintained. As a result, we now know that the scientific world is an integral part of the 'impure' world at large, and as such it is subject to political, economic, and other forces which 'prostitute' it for military or commercial gain.

Interference with the so-called ethos of science by social interests based on quite contrasting values, means that science can escape neither the impact of these different attitudes nor comparison with competing 'world-views'. Fortunately, this is by no means as unhealthy a situation as it may appear, and our purpose in this chapter is to examine the reasons why. First, we look at the influence on science of an ambient society which still exhibits an unjust and hierarchical organization; second, at a number of critical attitudes to the role of science as an intellectual sub-culture within the complex society; and finally, at some of the principal issues affecting the individual's relations with science and with the alternative, and sometimes conflicting, views of the world provided by religion. In so doing it is hoped that we may achieve a more balanced perspective, by viewing science from outside its own hallowed precincts. We shall see, for example, that science need not be the exclusive monoculture or secular religion that some advocates would thrust upon us, but may be treated as no more than one of several possible ideologies to which we, and society, may adhere. Once acknowledged, a pluralistic attitude of this kind might yet engender the compassion, tolerance and wisdom that together are necessary if science, operating in a deeply troubled modern world, is to fulfil the manifold aspirations of human kind.

THE SOCIAL STRATIFICATION OF SCIENCE

It is not just an idealized caricature of science that portrays it as a meritocracy. There is, in fact, a good deal of sound empirical evidence that, as a social institution, science is exceptional in allocating its rewards and determining its ranks very largely on the basis of what is perceived to be quality of performance. Although not unblemished, the scientific world appears to operate with a greater degree of independence from such factors as the nationality, race, sex or religion of its practitioners, than does any comparable enterprise. Where discrimination on one or other of these bases does occur, it is more often than not a reflection of attitudes in the wider society, attitudes which science, as a part of that society, cannot escape. But it is important to realize that the prevalence of discriminatory attitudes which rub off on science, need not in the strict sense be evidence of discrimination within the institution of science itself.

The situation prevailing inside the scientific establishment may at first appear paradoxical, for the result of its egalitarian, meritocratic structure, is extreme inequality and stratification. As we saw in chapter 6 only a small proportion of all scientists are sufficiently talented to make discoveries of first-rate importance, and it is to this elite that most of the major rewards accrue. Not surprisingly perhaps, this inherent system of stratification is reinforced by the progressive accumulation of advantage and disadvantage, as the most successful individuals gain access to the best facilities and the liveliest minds, while their less fortunate peers vegetate in unstimulating environments or drop out of the group of productive scientists altogether. And it is therefore by this means that the scientific world becomes polarized into a small group of 'haves' and a large group of 'have-nots'.

Could the mechanism responsible for this hierarchical organization within science, also account for the position of 'minority' groups, such as women or blacks, who traditionally have played a very small role in the scientific establishment? And could polarization *within* the scientific community somehow encourage the polarization that all too often results *from* science in its applications outside?

On the first question, it does seem plausible that the cycle of advantage and disadvantage that has long operated for 'white men of science' may very well operate also for blacks. (The position of women is considered in more detail in the next section.) Their position has been analogous to that of the under-privileged, provincial, whites, cut off from access to institutions of scientific excellence, the best mature scientists, and the best research facilities. But of course in addition to this, blacks have been the victims of historically entrenched racist attitudes which purported to find 'natural' differences of physiology and psychology sufficient to explain an apparent failure to compete. As we intimated above, there is abundant evidence here of social discrimination, but it is not clear that it is discrimination in science.

What is clear – alarmingly so to many 'non-white' writers – is that modern technological science blatantly discriminates in favour of the interests of the affluent North, at the expense of the impoverished South. This problem is seen as one of the inequitable distribution of the world's resources, coupled with an insidious process of economic colonization, both of them typically under the political guidance of the 'capitalist' system (see below the section on *The Counter-Culture*). There is, however, a contradiction in the impact of science and technology on the Third World (and, indeed, on minority groups within Western society). On the one hand it is recognized as a force of liberation from intolerable conditions of labour and disease, while on the other it is exploitive and 'imperialist', thrusting Western-style industrialization and Western values on to communities that still lack such elementary necessities as adequate food, clothing, shelter, medical and educational facilities, transport and employment.

This tendency has not only widened the gulf between the 'have' and the 'have-not' countries of the world but it has also accentuated inequalities within the poor nations themselves. This has been by creating a system that demands a small number of technical experts to control production, and a huge mass of the largely untrained to execute their plans. Power is thus concentrated in the hands of an elite – often consisting of 'honorary whites', who have been educated and encultured in the. rich world – so that democratic participation is undermined. Finally, it is said, the technologies that are imported into Third World countries typically produce luxury goods and trinkets to pacify the masses, and 'cash-crops' to swell the coffers of multinational companies. Meanwhile the real needs of the people are ignored, because to satisfy them – and their satisfaction has not been a technical or scientific problem for at least a generation – would be unprofitable.

Cynical and partial this interpretation may be, but it is one that speaks a good deal of truth. And, it need hardly be added, it is part of a problem of such magnitude that any long-term world stability appears inconceivable without its resolution. However, to propose even outline answers to such a problem would take us deep into controversial science policy studies, as well as into questions of ethics and politics which were considered in chapter 7 and are touched on again below. For the moment, it is perhaps wiser if we return to the question of stratification *within* science, by examining the prevalence of discriminatory attitudes to women.

Sex, Science and Society

Overwhelmingly, the 'great' scientists of history have been men. Until relatively recent times, the causes underlying this state of affairs were seldom sought, so powerful were the cultural and ideological pressures on women to conform to the norms of feminine docility and domestic servitude. With the rise of the feminist movement however, and the consequent widening of women's expectations to areas beyond the home and family, the traditional stereotypes and their social consequences have increasingly been challenged. Sociologists have made detailed investigations of the nature and extent of sexist discrimination generally, and by now there is a good deal of information specifically relating to science. Together with the scholarly literature, it must be said that there has also been a rapidly growing abundance of articles written by committed feminists. These are of variable quality, ranging as they do from penetrating analyses of particular problems, to the most general and embittered diatribes against the male-dominated society. Each article has its interest, although in extreme cases it must be feared that the concentration in a woman's pen of the sort of vitriolic aggression more conventionally associated with the worst of men, might well be counter-productive for the feminist movement as a whole.

At any rate, it is because the strength of feeling generated on such contentious issues as sex discrimination is seldom related directly to the quality of supporting evidence, that we are fortunate in having available an article and a book by the respected American scholar, Jonathan Cole. The first, 'Women in American science' (1975), was written with his equally distinguished compatriot, Harriet Zuckerman, and concerns the so-called triple penalty traditionally paid by women in their attempt to penetrate the scientific establishment. By this, the authors had in mind first, the cultural barriers which define science as an inappropriate career for women; second, the dismissive attitudes often adopted by male scientists to those women who *have* managed to enter the profession, attitudes which damn them as incapable of creative work; and third, the actual discrimination in terms of organizational barriers in the allocation of scientific opportunities and rewards. In Cole's book, *Fair Science* (1979), current beliefs about the treatment of women in science are examined further, the book's ambiguous title nicely encapsulating its analysis of the position of women in a community which prides itself on operating entirely equitable criteria of recognition.

The investigations suggested that the position of women in science might be determined by a combination of two potential factors, 'social and self-selection'. In this connection, the term social selection refers to discrimination against women by the institutions that employ scientists.

This of course is just the sort of prejudice that feminists fear and expect, but the evidence shows, at least in modern America, that such discrimination is largely illusory. Science, it seems, *is* substantially meritocratic.

Self-selection, on the other hand – although it is induced by social pressures of varying subtlety – is not operated by the universities and research establishments, but rather by women themselves. This process appears to be of considerable importance, and it partly explains why sex discrimination *within* science is comparatively rare, for most women simply opt out of their own accord. If they do compete within the scientific meritocracy, it often appears that women are confronted by a sort of 'Hobson's choice': either they can 'succeed by denying their femininity, or fail, confirming their inferiority'. In common with the exceptional black scientist, the woman who yearns for success must first attain an 'honorary' status which makes her effectively indistinguishable from the majority of her (male) peers. For example, to gain an independence and mobility comparable to that enjoyed by most of her married male colleagues, she will generally be obliged herself to remain unmarried. Even if she does remain single, it seems that the *higher* scientific honours are still more likely to elude her than similarly qualified men, this, in all probability, being evidence of a lingering sexist bias *within* science.

In view of the realities of socially induced self-selection, it is not unexpected that the values of the wider society are reflected in the sorts of science that women *do* penetrate. Roughly speaking, the 'harder' the discipline, the rarer are female practitioners; for example, of American PhD degrees granted in 1970, women comprised 3 per cent in physics and astronomy, 8 per cent in chemistry, 15 per cent in biology, 18 per cent in sociology and 24 per cent in psychology. Moreover, whereas with men the attainment of the doctorate tends to be regarded as no more than an initiation into the profession, with women it more often signifies the end of the research road. Having achieved it, women typically devote themselves more to non-research activities (for instance teaching), which not only command less prestige, but also result in a lower rate of production of the one commodity, knowledge, which the scientific community values. As a result they – together with unproductive male scientists – receive few rewards and are commonly to be found at the bottom of the scientific hierarchy.

Improvements have undoubtedly occurred for female scientists in recent years, hand-in-hand with the liberalization of public attitudes, particularly those concerning careers for married women. In simple statistical terms for instance, the percentage of science doctorates gained by females in the United States rose from 11 in 1969 to 23 in 1976 (although as we saw earlier, a smaller proportion of women actually continue in research afterwards). The feminist movement has been á major force in this advance, destroying many of the myths about women, creating a wider consciousness of their equal rights, and demanding affirmative and legislative action. Among its more radical writers, a good deal of emphasis has been placed on the dissonance between the alleged personality characteristics of 'typical' creative scientists (aggressive, competitive, even exhibitionist) and the passive, selfless, manipulating role for which most women have been socialized. In these terms, the very ideology of science is therefore identified as masculine, and fundamentally alien to the stereotyped female.

Two possible solutions to this dilemma have been suggested. Either women must be 'reformed' in the direction of the idealized male personality, their femininity submerged beneath a more assertive veneer – which, among other things, would raise a host of problems concerning the nature and distribution of parental responsibilities – or science itself must somehow be made more 'feminine'. For many people of both sexes, the first option seems defeatist, as well as impossible and undesirable. As to the second, it has been persuasively argued that since there is no shortage of positive attributes within the traditional feminine sphere, both scientific and human progress would achieve a more healthy equilibrium if these were accommodated as the natural comp-lements to the masculine attributes which dominate at the present day. It can hardly be doubted that any such transformation would have pro-found consequences for the structure and the direction of scientific practice.

CRITICAL SCIENCE STUDIES

Roots of Disenchantment
The various ideological elements identified within the complex entity called science (including sexism, and scientism itself) have led in recent times to an almost equally varied chorus of criticism. In spite of its growing dominance over our lives, by means of a uniquely efficient method and by its innumerable applications in technology, the scientific attitude to the world has not proved to be the panacea that some nineteenth-century optimists thought inevitable. Moreover, the price paid in terms of the decline of ancient and traditional attitudes has, according to the critics, been in many connections excessive. Urgently they advocate a reappraisal of fundamental values before, they say, it is too late.

Criticism of science, especially of its more mechanistic and material-istic conception of the world, is not new. Roots extend far back into history, at least as far as the scientific revolution of the sixteenth and seventeenth centuries, which is understood nowadays in terms of the struggle for dominance between three relatively distinct intellectual traditions or 'world views'. First there was the prodigious scheme derived from Aristotle of a divinely ordained universe drawn by analogy with the growth and decay of a living organism. This had reigned unchallenged as the doctrine of orthodoxy for two thousand years, being at once so compelling and so comprehensive that any alternative seemed inconceivable and unnecessary. Its eventual criticism and overthrow were possible only at a time of quite exceptional social and intellectual ferment: the wars, rebellions and constitutional crises, and the fertile radicalism in theology, politics and economics, all associated in some way perhaps with the rapid expansion of the known physical world.

At last, during this great revolutionary period which saw the translation of many ancient Greek texts, the complete works of Plato became available in the West, although still in the 'impure' form of Neoplatonism. These Neoplatonic teachings were welcomed by those humanist scholars who, intoxicated by the new atmosphere of discovery and progress, detected in the mystical and mathematical conception of the world a rival to Aristotle's 'organic' tradition. While the latter emphasized the rational and the empirical, the advocates of the alternative scheme were impressed by the 'magic' and mystery of the world, aspects with which the Platonic teaching seemed more in tune.

The third world-view took some time to gain self-assurance but was eventually to triumph over all. Its primary basis was to be found neither in Aristotle nor Plato, but in the also newly translated works of the Greek atomists and in the mechanics of Archimedes. These roots provided the basis for a quantitative, mechanistic conception of the universe which emerged in definitive form only late in the seventeenth century, but with such vigour that in a comparatively short time it attained a degree of supremacy that was incompatible with the survival of its competitors. Its dominance has been substantially unyielding down to the present day, although serious questioning did begin with the appearance of the New Physics in the early decades of the present century (chapter 5).

It was this atomistic-mechanistic philosophy which emerged as *the* modern scientific world view. Associated directly or indirectly (and, as recent research has shown, sometimes mistakenly) with the doctrines of Bacon, Descartes, Galileo, Boyle, Newton, Hobbes, Locke and Hume, it represented an attempt to detach science from religion by making it dependent solely upon experience of the physical world (empiricism). In so doing it had the effect of separating altogether factual knowledge

about the world from such seemingly intrinsic human values as good-
ness, truth and beauty. Hitherto, facts and values had been essentially
fused together as two aspects of one reality, in the Christian context at
least, with values (which were transcendental) as definitely the dominant partner. In the new expiricism, not only were facts and values
separated but, given the heavy emphasis on the importance of knowledge of physical phenomena, the position of abstract values became
distinctly secondary.

By the eighteenth century the empiricist philosophy was given further
impetus when France became the centre of 'progressive' thought in the
movement known as the Enlightenment. Although the founders of this
movement were by no means united in their attitude to science, such
influential figures as d'Alembert, Condorcet and Voltaire believed
fervently that the widespread application of Newtonian scientific
method would guarantee the securalization of society and bring the
fruits of reason and justice to all. Their fellow-countryman, Comte,
later developed on this foundation his 'positive' science of society,
which advocated the collection and correlation of facts and the
avoidance of all unverifiable speculation. Comte's work shared certain
basic assumptions with that of the nineteenth-century British utilitarians
Bentham and Mill, and the German materialist Marx and Engels, so
that by the early twentieth century the ground was fertile for the
emergence of the logical positivism of the Vienna Circle. This philosophy, as we saw previously, dealt with non-factual propositions such as
those of religion, by disregarding them as meaningless, while science
rested on supposedly verifiable facts (see chapter 4).

Despite the enormous influence of this orthodox scientific world view
(quantitative, mechanistic-materialistic, reductionist, 'tough-minded')
there has always flickered the flame of an alternative tradition which
traces its roots to the Neoplatonic revival of the seventeenth century.
It is this unorthodox conception (qualitative, mystical, holistic, 'tender-
minded') which has given rise to many critical science studies in our own
day. Broadly speaking, it has evolved through the works of such
'scientists' and philosophers as Copernicus, Kepler, Pascal, Leibniz,
Spinoza, Rousseau and Kant, reaching a significant flowering in the
early romantic movement of the eighteenth century. Rousseau in
particular propagated the idea that the social alienation of modern man
was a direct consequence of 'civilization', which he saw in turn as the
product of an empiricist world view. Only the 'noble savage' could have
access to genuine virtue.

Continuity with the Neoplatonic alternative tradition can be found in
the German nature-philosophy of the nineteenth century, associated
most notably with the poet and scientist Goethe. Here was a conception

of science and of the world in general, which was sympathetic to such 'alchemical' notions as natural magic, the universality of spirit, and the reflection of the macrocosm of the universe in the microcosm of man. Although nature-philosophy could not withstand the overpowering impulse of the harder experimental science, it lived on in the influential, though esoteric, doctrine known as theosophy ('wisdom about God'), which was a fusion of Neoplatonism with aspects of Eastern philosophy. For a direct philosophical root of the anti-empiricist, anti-science convictions of the present day counter-culture (see below) we should perhaps turn to the German mystic and educationalist, Rudolph Steiner, whose eclectic teachings represent a blending of Goethe with theosophy to produce a system which claims that the key to real understanding of the universe is within the human psyche. In some respects Steiner's English equivalent was the poet Coleridge, whose voluminous prose writings included a profound appreciation of the romantic ideal of science.

Before examining further the contemporary opposition to orthodox science, it is helpful if we first discuss in more general terms the uneasy relations which are said to exist within Western society between the 'new' culture represented by scientific method and scientific knowledge, and the more traditional culture of the classics, literature and the fine arts. An awareness of this tense dichotomy gives us some insight into how vague feelings of disenchantment may become hardened by isolation and incomprehension into an articulate and forthright criticism.

'The Two Cultures'

To some extent this conflict has been sharpened by the way in which the status of science has been progressively enhanced, especially in terms of its social utility, while at the same time the relevance of more traditional areas of scholarship has appeared to decline. In recent times a highly influential expositer of this problem has been the novelist C. P. Snow, who described himself as 'by training . . . a scientist: by vocation . . . a writer'. In his celebrated lecture, *The Two Cultures and the Scientific Revolution* (1959), Snow dealt openly, even provocatively, with what he saw as the main issues.

The word 'culture' has, as Snow pointed out, two distinct meanings, and both of them are applicable to what he wanted to say about science. In the first sense, culture is taken to mean 'intellectual development, development of the mind', so that we might say of an individual that he is 'cultured' or 'cultivated'. Snow's claim is that this meaning of the word applies equally to the development achieved by a scientist during his professional vocation, as it does to men of letters. However, neither the scientific system of mental development, nor the traditional literary method is, he claims, any longer adequate by itself to deal with the grave problems of the modern world.

Before we look at the problems Snow had in mind, it is important to mention the second sense in which he applied the word 'culture'. This is the anthropological usage which describes a group of people living in the same environment. Both scientists and literary intellectuals, Snow claimed, exist as members of isolated sub-cultures within the wider society. Each sub-culture has its own common assumptions, attitudes and standards and its common patterns of behaviour. This has promoted a separation of the two communities, not to say their mutual distrust, such that 'Greenwich Village [talks] precisely the same language as Chelsea, and both [have] about as much communication with M.I.T. as though the scientists spoke nothing but Tibetan'.

Snow himself was in an unusual position for he had a substantial foot in each camp. This may have contributed to the violent debate, and in particular to the personal abuse, which followed publication of his lecture. The reasons underlying these reactions may be an interesting subject for speculation, but the furore itself obscured to some extent the essential message that Snow intended to propagate. This message was not merely that there is an intellectual loss to our (Western) society as a result of the mutual incomprehension of the two cultures but, far more importantly, that there are dire consequences for the role of science in alleviating the immense and unnecessary suffering in the world. Until scientists feel the 'humanizing' influence of the traditional culture, and until the traditional intellectuals (in whose hands still lies most of the power to act) feel the *present* and accelerating ability of science to minimize the intolerable disparity between rich and poor, we shall continue to be incapable of applying science as the instrument for good that it so obviously could be. It is by its inability to produce 'whole people' that our present system fails so miserably to fulfil its duty to humankind.

The only solution to this tragic problem is, according to Snow, a fundamental re-shaping of our system of education. By means of this new system we are to eliminate the hostility between science and the arts and develop a superior, third culture, which can renew our intellectual and moral health. This may come from the influence of the social sciences, for they are to a substantial degree obliged to maintain contact with both of the other cultures (see also the comments in the introduction). In any event, we must delay no longer in introducing a form of education which can ensure that our young people 'are not ignorant of imaginative experience, both in the arts and in science, nor ignorant either of the endowment of applied science, or the remediable suffering of most of their fellow humans and of the responsibilities which, once they are seen, cannot be denied.'

For the best of humanitarian reasons Snow wished to spread the 'scientific revolution' around the world. He was referring, of course, to the revolution of twentieth century applied science, beginning roughly from 'the time when atomic particles were first made industrial use of. . . . The industrial society of electronics, atomic energy, automation, is in cardinal respects different in kind from any that has gone before, and will change the world much more.' To achieve this goal he saw both the need to train many more scientists in order to accelerate the growth of scientific knowledge, and the importance of enhancing their ethical sensibilities by exposing them to the world of literature.

Without doubting Snow's sincere motivation, a number of contemporary critics have questioned both the need for yet more scientific information, and the view that 'humanity' can be injected into science merely by greater familiarity on the part of its practitioners with the 'humane' arts. The two criticisms are very much part of the same general concern. While no one doubts that a legitimate goal of science is, and always has been, the expansion of knowledge, in our own time it seems to many that this has led to distortion by emphasizing quantity rather than quality. The search for new facts has thus become an end in itself, even though the information sought may be, in the words of the biologist Eric Ashby, nothing more than

> minuscule analyses of kitchen accounts in a medieval convent; the structure of beetles' wings – some beetle whose wings have not been studied before; the domestic life of an obscure Victorian poet; the respiration cycle of duckweeds. All, no doubt, interesting; all, in a way, at the frontier of knowledge, even though it is crawling along the frontier with a hand lens . . .

By Snow's own account, it does not require one additional scientific discovery to alleviate the world's suffering; lack of knowledge is no longer the problem. What is still lacking is a sure means of enabling humankind to discriminate between that knowledge which is worth having and that which is largely superfluous.

Snow evidently believed (and his most vituperative critic, F. R. Leavis, certainly believed) in the ability of music, or poetry, or painting to enhance human awareness and sensibility. This has been a recurring theme since at least the time of the Romantic movement, and there may well be something in it, despite those horrifying anecdotes concerning the Nazi concentration camp guards who returned home in the evening to listen to Mozart on the gramophone. But whatever the role of art and letters, what seems undeniably to be required before science can be orientated towards the good of man rather than to his destruction, is an enhancement of moral awareness, not merely of accumulated knowledge. There is little to indicate that scientific knowledge as we presently understand it can augment human virtue, and it is in this connection that Michael Polanyi's conception of 'superior knowledge' is of such importance. By this expression Polanyi means that knowledge, derived widely from each of the sub-cultures of human society, which is acceptable to all the 'great men' of each respective culture ('men to whose superiority we entrust ourselves, by trying to . . . follow their teachings and examples') as the most valuable. At the most fundamental level – or perhaps we should say at the level which transcends the realm of politics, personal and national competition, and so on – it should not, in fact, be difficult to achieve a consensus as to what knowledge and what action would most benefit mankind as a whole. At this level, there would be little disagreement about the pressing need to apply the best of scientific expertise to the international problems outlined by Snow. Ideally, the search for, and the application of, 'superior knowledge' might yet create a world of 'qualitative wholeness'.

The Counter-Culture: Anti-Science

Many of the fears about science voiced by members of the traditional literary culture have found renewed expression in an overt Anti-Science movement which originated during the 1960s in the belief that any such ideal as 'qualitative wholeness' was receding ever further from society's reach as a direct consequence of the forward march of science. One of the most influential books of the movement has been Theodore Roszak's *The Making of a Counter-Culture* (1968), which was presented not as a critique of the ideals or ethos of science as such, but rather of the 'technocracy', 'that social form in which an industrial society reaches the peak of its organised integration.' The anti-science movement is but a part of this wider counter-culture which questions many of the basic values of the Western life-style and interests itself in quasi-religious alternatives, including mysticism, meditation and psychodelic drugs. It particularly identifies reductionist thinking as something to be opposed. By reductionism it means that attitude of mind which interprets the needs of human society as a function of what can be technically achieved, and which then seeks to justify the technical achievement by an appeal to more basic scientific knowledge. Beyond this basic science there is no appeal, the clear implication being that science represents the ultimate in reliable knowledge. Roszak believes that science under the technocracy, has become *the* culture which dominates the lives of millions, and that failure of the counter-culture movement will 'leave nothing in store for us but what anti-utopians like Huxley and Orwell have forecast' (in *Brave New World* and *Nineteen Eighty-Four* respectively).

In the age of 'scientism' (chapter 7), Roszak realizes that many may read his book with little more than blank bewilderment. Yet there is no doubting its impact on critics and enthusiasts alike. Even if the Counter-Culture itself has lost some of its earlier momentum, the widespread discussion and the stream of publications leave no doubt that the anti-science movement continues to grow and evolve. In trying to understand the motivation behind it, the philosopher Stephen Toulmin (well known as a staunch advocate of disinterested science as a cultural activity) describes the image of science which is held up for rejection and repair.

> Science – or Technology . . . is depicted as an abstract, logical, mechanical, cold-blooded, generalizing activity, given over to collectivist policies and practices devised solely for their technical efficiency and without regard for their effect on all the heterogeneous flesh-and-blood individuals at the receiving end. For lack of personal insight, emotional imagination or a feeling for the specific impact of his activities on real individuals, the scientist goes his indifferent way, regarding his fellow-men simply as so much extra subject-matter for social and technological experimentation.

While many of these claims no doubt have the quality of caricature, Toulmin nevertheless asserts that they are not without some truth. They are, he argues, aimed at the profession of science 'for having delivered over to the temporal powers an enterprise which should be guided rather by perennial values'. By this he means, of course, that modern science has become inextricably linked with politics, while yet remaining unanswerable to those whose lives are affected by its actions.

A genuine problem therefore does exist and it is, in a word, that of how to make science more 'humane'. Some would argue that one of the more constructive reactions to the anti-science movement has been the growth of 'critical science', represented in the most general sense by an enhanced awareness of ecological balance (see 'Spaceship-Earth' in chapter 7). Toulmin himself sees the development of a humanitarian science in institutional terms, envisaging a framework of policy-making bodies the purpose of which is to defend the purely *human interests* of individuals in society against scientific and technological activities which threaten personal lives and happiness. Any proposal concerning 'human interests' however, takes us back to those difficult questions of quality, or moral judgement and, for many, of spiritual values, which were discussed earlier in this chapter and in chapter 7. It suggests to Roszak, for example, that we should be able to ask of every individual who would lead us, what his own talents have made of him as a *whole person*. Few might doubt that the genuine ideals of science – those unequivocally on the side of imagination, humanity and individual integrity –

could yet be realized if we were to 'reject the small souls who know only how to be correct, and cleave to the great who know how to be wise'.

The Marxian Critique

The sociology of science that we encountered in chapter 6 was almost entirely that of mainstream Western academia. At least until recently it accepted science as it found it, and was concerned only to analyse the details of its structure and function. But for Marxists, a mere interpretation of the world has never been enough; their real purpose is to change it.

An early instance of the systematic Marxian analysis of scientific growth was that presented by Boris Hessen, a Soviet physicist, at an international symposium held in London in 1931. In a famous lecture, Hessen argued that the mechanics of Isaac Newton – which he took to be a model for the emergence of modern science – developed as a direct consequence of the socio-economic forces of capitalism, specifically the needs of the rising bourgeoisie. His thesis, that science is in no sense an identifiably separate subsystem in society, but rather one aspect of the complex and closely interrelated processes of social development, was in the orthodox Marxian tradition. According to the Marxists this proved too much for orthodox Western sociologists, and provoked Merton's alternative explanation which emphasized social and religious values as opposed to material forces. Ultimately, as we have seen, this led to the notion of an ethos of science which was compatible with Protestantism and worthy of detailed examination independently of factors outside itself.

The Marxian critique of science is not so much 'externalist', however, as holistic. Science, it is argued, cannot be simply an autonomous institution which is acted upon by external forces, for rational attitudes are too much a part of human life for there to be truly separate social and scientific subsystems. Conventional Western sociology of science is therefore unsatisfactory in several important respects. In its emphasis on the work of the Great Men of science, whose contributions to the body of scientific knowledge are represented as virtually the whole of what is important, it is little more than a 'social psychology of the scientific elite'. In its refusal to acknowledge the essential unity of science and technology it is hopelessly unrepresentative, treating a small part as if it were the whole; science as a whole is not, and never has been in modern times, concerned only with the disinterested search for knowledge.

It is true that in ancient and feudal times there was an almost total separation between the work of scholars and of craftsmen, while scientific objectivity and rationalism were inhibited by the overpowering world view provided by religion. But with the Renaissance, these traditional barriers began to break down as the emerging class of capitalists saw that the innovations of science could be applied for commercial gain; thus began experimental science and the link with technology itself. After this, science did achieve gradual independence from religion but, say the Marxists, always found its inspiration in utility and the profit motive. By the later nineteenth century, industrialization demanded highly social forms of production and the interdependence of all aspects of human activity. In our own time the multidisciplinary research-and-development laboratory, which is the typical setting of twentieth-century science, has removed the last vestige of autonomy or non-utilitarianism from a science which is now wholly dependent upon the whims of political and economic policy.

The ideological bias which Marxists detect in Western sociology of science, also reveals its inadequacy as a theory of how progress occurs. For example, if the Mertonian norms of science can indeed be nurtured and maintained only in a 'modern liberal democratic society', but not in the more 'restrictive' societies of feudal Europe or Nazi Germany, how is it that there has been such progress in the Soviet Union, where science has never been claimed as autonomous or non-utilitarian? Equally, if Western science really does exhibit these ideal characteristics, how is it that so many American scientists have become dependent upon the military-industrial complex?

For most Marxists, of course, these questions are merely rhetorical. For them, science has developed according to 'laws of connections, between science and the other vital spheres of society, particularly of the material production processes as their fundamental basis'. Any scheme which fails to account for the impact of technological problems, and through them of economic forces, is doomed to irrelevance.

The Marxian alternative theory that is offered was pioneered in the West by J. D. Bernal's *The Social Function of Science* (1939), which reflected the influence of Hessen. Here there is a powerful evocation of the role of specific historical circumstances in promoting the economic conditions necessary to encourage scientific and technological innovation. Science is subservient to these historically determined social forces, and there can be no meaningful distinction between pure and applied science because the value of knowledge (including the cultural value) resides in its application. But while science and capitalism were at first interdependent – the latter encouraging science, which then generated profit and could be used as an instrument of power – there was nevertheless an inherent contradiction, because the ideological assumptions and distortions of capitalism enslaved science to the profit motive and prohibited its application for the good of humanity. Only a truly 'socialist' science, it is claimed, can ensure peace and plenty for all.

While Marx himself had realized that even in the nineteenth century science acted as an indispensable source of innovation in the capitalist system of production, and also as a force for social control, these twin themes have been particularly stressed by some contemporary writers. Thus Hilary and Stephen Rose speak of the 'incorporation of science into close correspondence with the technological and ideological needs of state and industry' (in both the Soviet and Western political systems). The critique of science is thus to be made in terms of its domination both of nature and of humanity alike. The gentlemanly 'cultivators of science' of the nineteenth century have become the oppressed and alienated 'scientific workers' of modern industry, unconcerned with, and ignorant of, the so-called scientific ethos, but preoccupied instead – just like other production workers – with conditions, security and pay. Their skills have become so specialized, their tasks defined by the machines they operate, that it is impossible for them to take any satisfaction from the end-products that result from such fragmented labour.

It seems that Francis Bacon's dream of the application of science for the improvement of 'man's estate' has now become a cynical nightmare in which human progress is reduced to technical progress. The vision of Marx and Engels of a time of harmonious equilibrium between nature and humanity has been replaced, first by the domination of nature by science, and then by the control of humanity by power. If we are to believe Marx, such developments are inherent in capitalist society. If we are influenced by the neo-Marxist philosopher Herbert Marcuse, whose *One Dimensional Man* (1964) presents a sombre analysis of society permeated by technological rationality, we may find little but pessimism in the ways of the modern world:

> Scientific management and scientific division of labour vastly increased the productivity of the economic, political and cultural enterprise. Result: the higher standard of living. At the same time and on the same ground, this rational enterprise produced a pattern of mind and behaviour which justified and absolved even the most destructive and oppressive features of the enterprise. Scientific-technical rationality and manipulation are welded together into new forms of social control.

Optimism is still to be found however among some Marxist commentators, even if their prescriptions remain vague. Since the capitalist system can be no more than temporary it follows, they argue, that its present misguidance of science, and its exploitation of nature and humanity, must in due course yield to a truly beneficent 'science for the people'.

The Problem of Accountability

When the apostles of the counter-culture speak of their alienation from science, and the radical Marxists refer to the alienation of scientists themselves, they typically have in mind a critique of science which emphasizes the sorts of research that threaten civilization. They point, for example, to the imbalance between the immense power of science to shape society and the powerlessness of society to control the development of science. In a provocative analysis of the social structure of science, J. R. Ravetz has gone further than this by identifying a contradictory element at the very core of the research process itself which, he argues, is largely responsible for isolating scientists from any sense of commitment to, or responsibility for, the fruits of their labour. It is quite conceivable that this element also underlies many of the conflicts between science and the wider society.

Ravetz's argument runs as follows: we would not for most purposes, hesitate to regard science as a fully-fledged profession. It provides full-time, paid employment for an especially learned group of individuals and, in common with the ancient professions of law, medicine and the Church, or with a more modern one such as engineering, it is represented by an institutionalized community with specific areas of interest, standards of performance and norms of behaviour. Furthermore, it offers a recognizably important service to the community. However, there is one vital respect in which science differs from other professions, for the individual scientist seldom plays the role of consultant specialist, whose professional calibre may be judged by the quality of his results. A relationship in which the specialist acts in the interests of a client and is personally responsible for the consequences of his actions, often characterizes the work of the lawyer, doctor, or cleric, who is likely to be thought accountable for other people's legal status, bodily or spiritual health. But while science as an institution is commonly identified as the villain in connection with problems such as pollution or the arms race, an individual scientist, because of the remoteness of his (academic) environment or the subservience of his (industrial) position, very rarely is. In most cases, the individual appears to be but one small cog in a gigantic and apparently unstoppable wheel.

This lack of individual accountability at the heart of an enterprise which collectively commands such power, is a deeply disturbing state of affairs. It represents indeed, 'a new twist on the old formula for corruption'; a contradiction apparently inherent in the research process, the ultimate consequences of which remain a matter for speculation. Should the overt acknowledgement of so dangerous a situation encourage us to believe that a fundamental restructuring of scientific ideals and institutions is still possible? Or is the contradiction simply too deep-seated a feature of modern technological science to justify anything so sanguine? As Ravetz himself says, only time will tell. Meanwhile, the identification of what has all the appearance of a malignant element at the core of research, is bound to enliven debate in the context of critical science studies. (Also see the section on Authority and Fraud in chapter 9.)

SCIENCE AND THE RELIGIOUS WORLD-VIEW

The reader who has proceeded thus far will now be well aware that modern science has at its disposal a body of knowledge so vast that no individual scientist can hope to master more than a tiny fraction of any one of its countless aspects. Yet he may equally feel that the accumulation of this knowledge has not been accompanied by a corresponding growth of fundamental understanding about the world. The knowledge has given an unprecedented degree of control over natural phenomena but it has done little to bring equanimity to the individual who, tiring perhaps of its ceaseless expansion, pauses momentarily to consider deeper questions which have to do with the purpose of his own life. While science can no doubt tell us something about everything, it seemingly can never tell us everything we would like to know. Indeed, it is because science has little or nothing to say on just those questions which often seem the most important, that many people, including many practising scientists, turn to more traditional ways of finding meaning and fulfilment in their personal lives.

To such people, it seems that the dominance of scientific empiricism, with its exclusive reliance upon 'facts', objectivity and reason, is destroying the essential unity of the human psyche by ruthlessly dividing knowledge from values. It is feared, moreover, that this division is ever widening, as modern technology endlessly bombards us with quantities of factual information that we can no longer handle; discrimination has been rendered impotent, as the accumulation of knowledge becomes an end in itself. Yet it has always been recognized, in both Western and Eastern philosophy, that the acts of knowing and of being are inseparably linked. That is, the quality of one's knowledge, the condition of one's mind, has an important bearing on the state of one's being, a realization that is sometimes encapsulated in Buddhism by the aphorism: 'You are not what you think you are, but what you think, you are.'

So ancient an expression of profound understanding finds little sympathy in a 'progressive' world that reduces wisdom to knowledge, and knowledge to isolated bits of information; traditional values to commercial imperatives and whole people to one-dimensional experts. As the philosopher Henryk Skolimowski puts it, we must 'remove many spurious dichotomies and distinctions, for they are often at the root of alienation in the present day world. Above all, we have to restore the unity of knowledge and values; we have to realise that wisdom or "enlightened knowledge" is the key to human *meaning*'.

Whether or not the problem of the meaning of human life is to be resolved by a renewal, or a re-exploration of old religious ideals, must ultimately be for the individual to decide. But it may at least be of some value if we finish by attempting to outline some of the questions which must be confronted by anyone who explores the reciprocal relations of science and religious belief.

Relations with Christian Theology

From the sixteenth and seventeenth centuries down to the present day, modern science has encountered the established authority of the Christian religion both as a hindrance and as a stimulus. Although the inspirational role of religious attitudes and Church doctrine must never be underrated, it is to the 'conflict' between science and religion that most attention has been directed. During the nineteenth century there is little doubt that this was the aspect most discussed, as for example in famous books with highly explicit titles such as *History of the Conflict between Religion and Science* (1875) by J. W. Draper, and even *A History of the Warfare of Science with Theology in Christendom* (1895) by A. D. White. In our own day religion has become less concerned with 'facts', while at the same time science has become a good deal more modest in its claim to account for the world. Yet many thinking people, some of whom even retain a nominal allegiance to the Church, no longer find it possible to believe that religious ideas are derived supernaturally. A widespread contemporary view is that it was man who created the concept of god, not God who created man. The main reason behind this change is undoubtedly the influence of science during the last three hundred years.

At the heart of the conflict in its traditional form has been the contrast between the absolute, unchanging authority of divinely revealed religious doctrines and the ceaselessly evolving world-view derived from science. Thus one of the most famous confrontations of all, that between Galileo and the Catholic Church, turned on the supposed threat posed by Galileo's *new* observational data to the long-established dogmas of Aristotelian cosmology and the biblical narrative. More than two hundred years after this episode it was to be the theory of evolution which generated violent controversy, and this again centred around the challenge to the traditional Genesis account of the creation of species, in particular the creation of man, wrought by the *new* evidence from geology and natural history.

The general problem posed for the religious view by such advances in science was that as more and more *natural* causes of wordly effects came to be uncovered, so there was less and less need to postulate supernatural causes, oŕ divine intervention. When 'battles' were fought, there seemed to be a marked tendency for them to be lost by the Church, while even for those effects as yet unexplained, there remained the danger that new scientific discoveries might serve only to undermine the authority of religion further if, so to speak, God were dragged in prematurely to 'stop the gaps'.

The answer to this problem came in several forms. Some fundamentalist theologians simply dug themselves in behind an impenetrable barrier of scriptural literalism; for them, every word in the Bible was taken as revealed directly by a personal God, and anything that controverted it was accordingly untrue. At the other extreme the 'modernists' of the later nineteenth century saw in evolution an impersonal cosmic force which seemed to them evidence enough of a Divinely ordained natural law of 'progress'. For such progressive thinkers, God and man together became a part of Nature, such that *any* new scientific knowledge could be construed as supportive of their position; the Bible therefore ceased to be party to any controversy with science. This is not to say however that the Bible was of no significance, for it became, particularly for theologians of a middle ground, the source, not of a literally true account of the world, but of the revelation of God in the person of Christ.

This notion of a truth revealed by supernatural communication had the effect of isolating science and religious belief by defining their respective methods and subject matter as having reference to wholly different 'worlds'. This construction permits subtle ideas about different 'levels of truth' and different meanings of the word 'explanation', which are complementary but not contradictory. While science explains the existence of different forms of organism in terms of evolutionary theory and, ultimately, in terms of the emergence of life from inanimate matter, it has nothing to say on the question of existence *per se*. Existence itself is the concern of religious explanation. Similarly, the scientific explanation of the origin of our universe in terms, for example, of the 'Big Bang' theory, begs the question of the source and status of the matter or energy available to explode in the first place. On this question, again, science has nothing to say.

In general terms this theological position, shared as it is by many Protestant and Catholic thinkers alike, and favoured as a means of minimizing the area upon which science and religion *can* debate, argues that the whole problem of creation is, for science, one that is treated on the level only of the finite and temporal relations between cause and effect. For religion, by contrast, the problem of creation is fundamentally about the meaning of existence, in particular the meaning of the *present* relations between God and the world. The insights derived from revelation are seen as impregnable to the discoveries of science because they refer to altogether different levels of causality, primary and secondary causality respectively.

Other philosophers and theologians too have wished to resolve any conflict between science and religion by separating the two into distinct realms of experience. For example, the existentialists dissociate a sphere of selfhood, or subjective involvement, in which God alone operates, from a sphere of nature, or objective detachment, familiar to the scientist. On the other hand, the linguistic philosophers acknowledge the legitimacy of both science and religion, yet understand them as totally distinct *languages*, the one concerned with prediction and control in nature, the other with the endorsement of a particular way of life.

In the last quarter of a century there has been something of a move towards re-uniting religion and science. Both in terms of method and of content it has been argued that the two have more in common than has usually been supposed. The minority of theologians who encouraged this coming-together included those influenced by the reassessment of the concept of 'scientific truth' by philosophers such as Thomas Kuhn and Michael Polanyi, for whom there is a far greater involvement on the part of the individual scientist than has hitherto been acknowledged (see, for example, the section on Thomas Kuhn in chapter 4). The theory-laden nature of scientific observations, the powerful psychological commitment to paradigms, and the view of science as consisting ultimately of *Personal Knowledge*, all helped to erode the once-rigid separation of 'objective' or 'positive' science from 'subjective' or 'normative' religious belief. Nowadays the simplistic distinction drawn by nineteenth-century rationalists between factual science and fictional religion is unlikely to be taken seriously even by the most hard-bitten, atheistic, modern scientist assuming, of course, that he is *aware* of these recent philosophical trends. No matter how dismissive he may be towards the status of revealed truth, he will nevertheless be obliged to recognize the relativistic nature of scientific knowledge, a knowledge which is no longer described in terms of infallible 'facts' but rather in terms of fallible observations and models. None of these models is regarded as *the* definitive account of reality but all are treated as no more than partial accounts and temporary approximations. It is not by any means necessarily the case that science supports religion, but rather that science appears no longer to be in a position to indicate that religion is untrue.

In this general retreat from a positivistic interpretation of scientific knowledge, some scientists and theologians have found a role for 'belief' in the methods of science as well as in the methods of religion. Among the implicit assumptions of science are such notions as the intelligibility, uniformity, orderliness and 'simplicity' of nature. Science does not proceed by accumulating facts but rather by the creative involvement of the imagination, which interprets empirical experience according to pre-conceived ideals in much the same way as it does with religious experience. When, to these basic parallels, are added the similarities of attitude found within both the scientific and religious communities, attitudes such as honesty, co-operation and universalism, it is, according to this view, perfectly possible to envisage a truly comprehensive world-view (but perhaps not an 'objective' one) based upon a totality of experience derived from both complementary sources.

As to the question of similarities of subject matter, it has been the enormously fruitful theme of evolution which, perhaps more than any other scientific theory, has encouraged a fully comprehensive view. An attempted synthesis between the cosmological and evolutionary ideas of science on the one hand and the insights of revelation and mystical experience on the other, has been performed substantially upon the basis of the writings of the mathematician and philosopher A. N. Whitehead and the Jesuit paleontologist Pierre Teilhard de Chardin (see also the section on *The Ideas of Biology* in chapter 5). According to their vision, the world is an interacting whole, still in an incomplete state of becoming. There are various levels of reality within this whole, the cosmic processes of evolution 'directing' the emergence of successively higher levels, from matter to life, from life to conscious-ness and from conscious thought to society, but without absolute breaks in the vast continuum. Thus man is seen as an integral part of nature whose unique level of organization gives him alone access to the 'within of things' and by which he may attain some remote comprehension of what experience might be like at lower levels. The idea of the evolution of greater awareness acknowledges a role for chance and indeterminacy in nature, because with increasing awareness comes ever greater freedom, the higher forms of being having to react selectively or voluntarily to the many potentialities that the world presents to them. The process of creation is thus seen as continuous through this recurring novelty, and it is by a profounder appreciation of evolutionary processes that the God known from religious experience may be better understood.

Belief in the Existence of God

In the absence of some deeply significant religious experience – that is, of something interpreted by the individual himself as a direct revelation of God – can belief in a supernatural Power still be consistent with the knowledge possessed by a modern scientist? This initial qualification is an important one, for mystical experience of 'God's presence', of the 'Ultimate Reality' or of the 'One', as reported by the holy men of all the world's religions and by innumerable lay people as well, seems without question to be an event which can produce an unequalled sense of peace and purpose and which, unlike any other, can change lives. Simply because those who have not had such experiences can quite legitimately question their authenticity (how, for example, are they to be distinguished unequivocally from apparently analoguous experiences which may be induced by chemicals in the sane, or from the compelling 'voices' which we think it right to eliminate by chemicals in the insane?) this does not influence in any way the conviction of those who have. The mystic can no more *prove* to the sceptic that he has indeed experienced God, than the sceptic can prove to the mystic that he has not. For the mystic himself, the directness of his experience will have transcended the level of the rational mind; unlike the claim of science, his claim to knowledge will be absolute.

Most contemporary scientists are more likely to feel closer to the sceptic than to the guru or the saint. They will accordingly wish to rely more on their rational faculties than upon any process that may smack of 'wish-fulfilment' and will be interested in those rational arguments for the existence of God which are not incompatible with scientific knowledge. There are of course many possible candidates for such arguments, but it will suffice if we very briefly mention the three most famous ones.

Perhaps the oldest of all is that commonly known as the cosmological argument which we touched on above in connection with the concept of revealed truth. In its Aristotelian form, the assumption is first made that every effect must have a cause. On this basis it is then argued that the effect we know as the universe must also have a cause. What has to be explained is thus the *existence* of the universe or, indeed, the existence of anything. The explanation for the existence of the universe is that it is the effect of the First Cause. This First Cause must be God, or, as the argument summarizes it, because the universe exists, God must exist.

Another form of reasoning which stems at least from medieval times, is that known as the ontological argument. This tries to show that an examination of the idea of the existence of God must lead to the conclusion that he does in fact exist. The critical step here is to define

God as 'something than which nothing greater can be conceived'. On this basis it is then argued that if something exists outside of the human mind in addition to being a concept of the mind, it must be greater than something which is a concept of the mind only. God *must* therefore exist outside as well as inside the mind, for otherwise he would not be the highest possible concept of contemplation. Because we can conceive of God's existence, it is claimed, God must exist.

The third of the classical cases, and by far the most popular, is represented by the teleological argument, commonly known as the argument from (or to) design. This is also of great antiquity, although it reached a peak of controversy in the eighteenth and nineteenth centuries. It is perhaps more an assertion than an argument, claiming in essence that the extraordinary complexity and organized pattern of things in the world, and especially of living things, can be explained only in terms of the work of an intelligent Designer. This Designer must be God.

Needless to say, each of these arguments has had its advocates as well as its critics. On the philosophical level, Kant is thought by many people to have disposed of the cosmological argument by showing it to be dependent upon the ontological argument which he also showed to be logically invalid. The details of Kant's logic, and the possible counter-arguments to it, are beyond our present concern. For our purpose, it is sufficient to say that, for Kant, rejection of the ontological argument was based on the distinction between what exists in reality (outside of the mind) and what exists merely in thought. His position is commonly paraphrased by the expression: 'You can't define things into existence.' The argument consists, he said, of two independent propositions; first that God is the highest possible concept of contemplation, and second that God must exist. But the second is not a logical consequence of the first because existence cannot be a property of God in the way that supremacy or perfection might be. Existence is presupposed by the attribution of the other properties, while the latter are taken for granted once existence is assumed. Thus while it *may* be false to say that God does not exist, it is nevertheless conceivably the case. According to Kant, the ontological argument does not *prove* God's existence.

On the scientific level it would seem that very little could be said about the ontological argument either way. The cosmological argument is in much the same position (although it may be of interest to some to recall that the notion of cause and effect seems to be of uncertain status in the world of particle physics). But generally, one's own views as to the argument's plausibility – what one understands by the 'creation' of the universe – will depend in turn upon the meaning that one attaches to the concept of 'explanation' in this context.

On the third argument, that from design, scientific evidence might appear to have a more immediate impact, although judgements as to the significance of the evidence have varied widely. Many scientists have found in their observations of nature support for their pre-existing belief in God, while others, particularly if they have emphasized the wastage and suffering of the natural world, have been confirmed in their atheism. The eighteenth-century doctrine of Deism came close to replacing the orthodox Christian concept of God as immanent in nature, springing as it did from criticism of the teleological argument by Hume and from an over-simplification and misinterpretation of the work of Newton. It was argued that while evidence of design might be held to favour the existence of God, such evidence gave no reason for believing in Almighty God. Accordingly God was, as it were, pushed into the background as the Creator of the universe, the Maker of its clockwork mechanism, but He was not regarded any longer as actively involved in its workings. In the nineteenth century, the impact of Darwin undermined the traditional teleological argument by attributing the structural and functional characteristics of organisms to a prolonged process of variation and natural selection. In the event, the only defence for many seemed to be a withdrawal into Deism and this, it has been argued, appears eventually to have become Darwin's own position.

For those who wish for a clear answer to our initial question – is it possible for the modern scientist to believe in God? – the problem with Deism is that in the final analysis it appears to fall back upon the cosmological argument; God becomes no more than the First Cause. But the cosmological argument is at best shaky because it depends upon the ontological argument which, if we are to believe Kant and most orthodox philosophers since his time, has to be rejected as logically unsound. Therefore it seems that an impressive weight of opinion indicates that there is little *rational* basis for a belief in God, and the scientist who has not experienced personal revelation will have only one other justification for holding such a belief, namely that of faith.

Certain acts of faith, however, are not the monopoly of those willing and happy to describe themselves as the 'faithful'. As was said above, all scientists work with limited preconceptions which can hardly be 'rationalized'; we saw in chapter 1 that an 'intuitive leap' is necessary in formulating a scientific view of the world; and in chapter 2 it was largely taken for granted that the rules of deduction do, indeed, allow us to make the inferences that we claim as valid. But although fundamental acts of faith of this kind may be a necessary prerequisite for living in the day-to-day world, faith in God does seem to be of a different order. All that need be said about faith at this level is that it is experienced by many people, including many scientists, as *the* most profound motivation without which their lives would be devoid of meaning or purpose. The fact that it may be experienced despite, rather than because of, purely rational arguments as to the existence of God, seems only to add the conviction that faith operates in a realm altogether separate from reason, a realm which, like that of revelation itself, transcends the logical faculties of the mind. The compelling nature of faith tends to be understood by the believer in terms of 'Divine Grace' and for the most devout this may remove altogether the last vestige of uncertainty. For the rest however, beliefs derived by faith may seem no more than the fulfilment of deep-seated psychological insecurities – often induced by fear of death or a sense of the purposelessness of the world – and, for those who are yet unwilling to take the ultimate step to outright disbelief, philosophic doubt will be the only position that it seems possible to entertain.

Whatever the influence of rational arguments, it seems that the human proclivity for 'irrational' faith may, at the most profound level, render impotent the power of scientific evidence to persuade. In the full knowledge of the significance of science, it still remains perfectly legitimate to hold a theistic, atheistic or agnostic position according to one's own personal convictions. Whatever psychological interpretation one may read into the position of others, it is at least well to remember that others may equally read significance into the position taken by oneself.

9

The New Sociology of Science

This chapter is in many respects the antithesis of what has gone before. Its purpose is to challenge the reader to reassess his or her attitude to the established philosophies and sociologies of science outlined earlier. Because it is devoted largely to developments of the last ten or fifteen years, there is a sense in which it should perhaps have come at the very beginning. Yet to begin an introductory book with a viewpoint that may appear radical and disorientating, might mean that the rest – which does, after all, summarize an impressive body of more traditional theories – would not be read at all! It is not the intention of the chapter to suggest that the traditional theories are without importance as a basis for interdisciplinary science studies; even if they increasingly appear to provide an unsatisfactory, perhaps a false, picture of science, they still represent a worthy 'establishment' view without which the new sociology would have had nothing to react against. In considering what follows, it should therefore be borne in mind that the current trend towards a wholesale debunking of the conventional ideal of science – as the noble pursuit of objective truth – together with the substitution of an 'instrumental' model of scientific knowledge as nothing more than a product of social forces, is no doubt itself in some ways a reflection (and conceivably even a fleeting, partial, or distorted reflection) of its time; we might speculate that it is analogous to modern-day 'relaxed' attitudes in literature, music and the other arts, and perhaps awaits analysis by a future sociology of the sociology of science! Although the new trend is not without its critics, it is now of unquestionable importance, firmly established, and highly influential; we can no longer afford to ignore it.

SCIENCE STUDIES: HISTORY, PHILOSOPHY AND SOCIOLOGY

The therapeutic value of Kuhn's or Feyerabend's philosophies of science lay in their providing for science a more human face. They derived from a study of history and did not attempt to prescribe what scientists *ought* to do. By identifying elements of irrationality they were clearly theories 'of their time', resonating widely (we might say) with contemporary currents in world affairs. The trouble was that Kuhn's major insight illuminated *The Structure of Scientific Revolutions* far less clearly than it did the introverted, conformist nature of 'normal science' itself. There can be little doubt that a conservative, rational, evolutionary, model of scientific progress was that approved by most practising scientists at the time of Kuhn's first edition (1962). It was the model they grew up with, and the one with which they felt most at ease. With some exceptions, this is probably still true today, even though the informed popular perception of science may be rather more open to such ideas as internal conflict and revolutionary change.

In order to understand something of why it is that scientists and lay people alike adhere to an 'old-fashioned' image of science so different from that of many scholars who study it professionally, we must briefly examine, first, the historical relations of science and philosophy, and second, the influential stereotype that was inherited from traditional interpretations of the scientific revolution.

Philosophy and the Scientists

The painful truth is that the great majority of modern scientists are almost entirely isolated from contemporary trends in philosophy and, if they have any view at all, it is to dismiss it as irrelevant to the practice of science. (Or at least they have come to regard that branch of philosophy concerned with the way of knowing – epistemology – as irrelevant; but we saw in chapter 7 that there is increasing evidence that they are becoming more concerned with another branch, ethics.) It is important that we realize that this represents a complete break with the past, for not only the Greeks, but many 'natural philosophers' from the sixteenth and seventeenth centuries, up to and including the physicists responsible for the revolutions of quantum mechanics and relativity theory in the early twentieth, were convinced that the two disciplines were necessarily and inextricably linked as parts of the overall structure of human knowledge about the world. Copernicus, Kepler, Galileo, Newton, Leibniz, Descartes, Kant, Darwin, Whewell, Heisenberg, Bohr, Einstein, and countless others, all subscribed to this tradition and all contributed richly to it.

It is a curious irony that it was the work of one of these eminent scientist-philosophers, Ernst Mach, that was chiefly responsible for the separation of science from philosophy. Mach believed that the natural sciences, and in particular his own discipline, physics, had, by the 1880s become altogether too dependent upon unobservable entities such as space, time and energy. All terms without empirical foundation should, he argued, be excluded, leaving only a 'positive' science in the spirit of Comte, freed of vague metaphysical concepts. Mach's empiricist definition of meaning thus made observation itself the essence of scientific practice, and observational terms the sole basis of its theory. In its original form positivism therefore raised some extremely difficult problems for mathematics, the very heart of theoretical physics, since its terms obviously had no empirical base. This bizarre situation was resolved, however, by elements of compromise which accepted mathematical propositions as meaningful by virtue of their association with the practices of scientists. In any case, it was argued, they were not indispensable to science, for they could in principle be reduced to series of statements containing only empirically meaningful terms. When Rudolph Carnap produced what was at first regarded as the definitive synthesis of (logical) positivism in 1923, it was believed by enthusiastic advocates that scientific thought could now at last be reformed by subjecting it to the most rigorous analysis by symbolic logic. Such terms as space and time, for example, would no longer be used in the old vague sense, but only in the context of observations, as 'measurements with a ruler' or 'measurements with a clock'. The idea of velocity would then be derived simply as 'space divided by time' and that of acceleration by further division. On the other hand, any term that could not be drawn logically from empirical foundations would be rejected from science.

Unfortunately for the positivists (as was noted briefly on page 53), grave internal (logical) problems were soon exposed, leading to seemingly endless complex and highly esoteric debates which could be of interest to scientists in only the remotest sense, if at all. Indeed, it has often been noted that the obsessive analysis of logical technicalities introduced into philosophy of science problems that did not reside within actual science at all. When deficiencies were exposed, they therefore tended to be those inherent in logical theory rather than in the structure of scientific thought itself. Not surprisingly, this tendency succeeded in alienating the vast majority of scientists from any interest in philosophy, to the mutual impoverishment of each discipline. Just as the philosophers, with their single-minded fixation upon physics, acted as if it were the whole of science, so too the scientists, for reasons all too easy to understand, have come to treat positivism (including falsificationism which, of course, stems from the same tradition) as if it were the only philosophy of science.

Rightly or wrongly then, and unlike his predecessors, the modern scientist has learnt to ignore philosophy. He believes that he 'knows' what scientific method is, and how he should go about it. He is content to be on automatic pilot and, like the centipede in the rhyme,

. . . was happy quite until the toad began:
Now pray, which leg goes after which?
Which wrought his mind to such a pitch,
He lay distracted in a ditch,
Reflecting how to run.

For many scientists, philosophy of science has become a 'debilitating befuddlement', a view that is fully justified even by an eminent contemporary philosopher of science, Hilary Putnam. All of the great schools of philosophy of science, he says, have ended in failure. 'Rational reconstruction' of the world according to some fool-proof mechanical procedure – the great hope of the early positivists – is now widely seen as unattainable, for at critical points it is always necessary to fall back upon instinctive human judgements of what is 'reasonable'. A modern example of this comes from a paper by the Nobel Prize winning physicist Sheldon Glashow (with H. Georgi). 'Our hypothesis may be wrong and our speculations idle', he writes, 'but the uniqueness and simplicity of our scheme are *reasons enough* that it be taken seriously'. In strictly formal terms this would have seemed outrageous, but in the pragmatic, operational world of modern science it is increasingly the day-to-day norm. Of course, it is not that rational argument and justification have been abandoned, but rather that a more modest appraisal is now made of what they can be expected to achieve. In his attempt to understand the world, today's scientist hopes that by some little-understood process of communal reasonableness, his own interpretations of phenomena will converge upon those of others. Moreover, he no longer expects to solve problems in any final sense, but is satisfied merely to contribute to gradually emerging temporary solutions which will themselves determine the future course of still better work. Putnam draws a vivid model of the new methodology in terms of a fleet of boats:

> The people in each boat are trying to reconstruct their own boat without modifying it so much at any one time that the boat sinks. . . . In addition, people are passing supplies and tools from one boat to another and shouting advice and encouragement (or discouragement) to each other. Finally, people sometimes decide they do not like the boat they are in and move to a different boat altogether. (And sometimes a boat sinks or is abandoned.) It is all a bit chaotic; but since it is a fleet, no one is ever totally out of signalling distance from all the other boats. We are . . . invited to engage in a truly human dialogue; one which combines collectivity with individual responsibility.

This picture – one of independent creativity disciplined by accountability to shared experience – brings us back full circle to writers such as Kuhn and Feyerabend. Their initial challenge to the positivist position is now developing in a new, and more realistic direction which construes philosophy of science itself in much more empirical (or scientific) terms. As the influence of the discredited *a priori* philosophies has faded, so the impact of historical investigations has been more felt. Scientists no longer find themselves told how they *should* behave but, instead, discover a growing interest by philosophers (and particularly sociologists) in their actual behaviour. Gone are the logical prescriptions and unintelligible linguistics of an alien culture; at last there are unmistakeable signs of a new respect for scientists, not as impersonal automata, but simply as human individuals participating in a culture common to all.

Science: Portraits of its Past and Present

It may be that the decline of the traditional epistemological concerns of philosophers of science and their replacement by a pragmatic empiricism, may yet succeed in encouraging scientists themselves to explore intellectual territory beyond the narrow technicalities of particular disciplines. Any such optimism will be hard to justify, however, until the traditional, and still widely influential, 'heroic' vision of science is overthrown.

The traditional view was heroic (and romantic) because it cast natural philosophers as selfless truth-seekers, valiantly battling against the entrenched odds of ignorance and prejudice. It was thought that truth, in the shape of objective facts about the world, must surely speak directly to unbiased observers, and give access to reality itself. Some such conception was implicit within most accounts of scientific progress during the sixteenth and seventeenth centuries, for example in those given us by the distinguished historians Herbert Butterfield (in his book *The Origins of Modern Science*, 1949) and A. Rupert Hall (*The Scientific Revolution*, 1954). Science was a noble activity, set apart and above the mundane tide of general culture, the scientific revolution having transformed and liberated human thought by substituting mechanism and mathematics for magic and mystery. By emphasizing the autonomy and uniqueness of science, these scholars created the impression of an enterprise which, though empirical, indeed increasingly experimental and utilitarian, was nevertheless at heart intellectual in nature, an enterprise 'whose object it [was] to gain some comprehension of the cosmos in terms which [were] in the last resort philosophical'. If some of the philosophical notions that science appeared to affirm – determinacy, purposelessness, infinity – seemed fundamentally alien in human terms, this was simply the price of the new maturity, and it had to be unflinchingly accepted by all who valued the advancement of objective knowledge by a method uncorrupted by pre-conception or wish-fulfilment.

It seems that we no longer believe in romantic heroes or idealised models of action. The new history of science, influenced as it is by the new empirical philosophy, has been marked even more by recent trends in the sociology of knowledge. No longer are 'facts' thought by themselves to provide (even the most unbiased) observers with an absolute, unambiguous version of reality; the new model holds that both 'truth' and rationality are more or less relative, depending upon social contexts and social processes which seek to achieve consensus. According to this view, scientific knowledge is not, after all, remote and unimpeachable but, like any other cultural product, is meet for sociological analysis. The consensus that is achieved in well-established areas of research, is therefore thought more likely to reflect the social organization of (normal) science than the definitive structure of reality. On the other hand, within pioneering research areas, or given radical new discoveries, unanimity may simply not be attained, or may break down (extraordinary science), giving rise to considerable confusion and not uncommonly to bitter personal disputes. Although we must never forget that the great majority of science is undoubtedly 'normal', it is clearly in the context of extraordinary science that the traditional model is unable to cope and the new sociology most anxious to intervene.

What this intervention has achieved is to make scientists much more like other creative people, helping to shape, and being shaped by, a host of wider social forces. The Great Men of the earlier tradition, working, so it seemed, in glorious isolation from the rest of culture, are seen in the light of recent scholarship to have been, after all, truly men of their time. Thus the heliocentric scheme of the universe was advanced by Copernicus in opposition to Ptolemy's earth-centred model, not because it was markedly superior in predicting the movements of the planets, but rather, it appears, because its aesthetic qualities of harmony, simplicity and purity (of circular motion) better reflected the ancient metaphysical doctrines to which he subscribed. Even with Newton (the super-hero of the scientific revolution) much the same applies, for his near-heretical acceptance of such ideas as the void (absence of matter) and action at a distance (gravity), was at least in part possible because his neo-Platonic philosophy valued the spiritual over the material and gave credence to 'magical' notions like affinity or sympathy. Early modern science was thus unquestionably shaped by occult influences, such that we can say without hesitation that chemistry emerged to a significant degree from alchemy, astronomy from astrology, however non- or pseudo-scientific such disciplines may now appear. As Roy Porter puts it: 'Science proclaimed its emancipation from the tutelage of magic, morals and metaphysics. But these manoeuvres remain most murky, and science absorbed more of these than it admitted'. If, under the traditional interpretation, we have thought at length about man's place in nature, 'we must now stand this explanation on its head, or rather its feet, and contemplate nature's place in man'.

THE SOCIOLOGY OF SCIENTIFIC KNOWLEDGE

Anthropologists have always tended to explain the beliefs of human communities by reference to aspects of their social organization. Similarly, sociologists interested in specialised bodies of knowledge – such as systems of religion, morality or politics – have been concerned to show how they are related to wider cultural aspects of the societies producing them. Yet this sort of analysis has, until recently, almost never been applied to that great body of knowledge that so characterizes modern (Western) society, namely natural science. Like the philosophers and historians, sociologists have assumed that science is exceptional in being very largely autonomous. In the following section we must briefly re-iterate the reasons for this past neglect of the sociology of scientific knowledge (as distinct from the sociology of scientists or of the scientific community) and then examine the great changes that have occurred in recent years in tandem with those in the history and philosophy of science.

The Traditional View

The essential philosophical question that underlies the sociology of knowledge concerns the degree to which the validity of thought depends upon social factors in its creation. Until recently, scientific knowledge was regarded as 'positive' because of the alleged objectivity of its methods, whereas in social or historical thought 'truth' was no more than the product of consensus. The unique epistemological status of scientific knowledge was therefore seen as excluding any sociological analysis of its origins – for social influences were simply not involved – and as restricting philosophers to the logical details of its justification.

Even before the era of logical positivism, the nineteenth century thinkers were at best uncertain about including natural science within any sociology of knowledge. Marx concentrated on the comparison of ideologies and on the social factors that led to the production and acceptance of different kinds of knowledge by different societies. Although we have seen that he influenced others (for example Boris Hessen) in regarding science as a social creation, indeed as an ideology in its own right, it is accepted by most commentators that within a given research specialty, knowledge-claims were seen as essentially non-ideological (in any wider sense). Social forces thus determined the direction of science, but not its internal, cognitive activities. For Durkheim, similarly, science was directed by changes in the structure of society, but the distinctive features of the scientific community increasingly guaranteed its liberation from social constraints. Sociological analysis of scientific knowledge was accordingly impossible, for to just the extent that it was influenced by social factors was it also not genuinely scientific.

Writing in the twentieth century, but in the Marxist tradition, Karl Mannheim was also convinced that scientific knowledge was not amenable to sociological investigations. The phenomena of the natural world were 'timeless and static' and could be studied only by impartial observation and accurate measurement. On the other hand, cultural phenomena could be analysed by interpreting the participants' meanings, which in turn meant understanding the values of their social groups. The distinction between scientific knowledge and social or historical knowledge was therefore that the former could be absolute, the latter merely relative to particular social contexts.

This distinction echoed that of the logical positivists by re-enforcing the notion that only questions concerning the validity of knowledge ('justification') were legitimate in science, those having to do with its social origins ('discovery') being inapplicable. Significantly, in his influential 'Paradigm for the Sociology of Knowledge', Robert Merton (1957) concentrated almost entirely on questions of the latter kind, thus indicating unambiguously his own belief that the operation of the 'ethos' of science ruled any sociology of scientific knowledge out of court.

Science and its Social Resources

The empirical survey of scientists and technologists undertaken by N. D. Ellis in the late 1960s (see chapter 6) called seriously into question the very idea of explaining science in terms of its socialization into 'norms and values'. Could this long-revered model of action be wrong after all? If so, then it would surely bring to the surface a controversy which, by the early 1970s, was beginning to rage furiously around the academic world. It was an important controversy because it highlighted wider discrepancies between the idealized accounts of science beloved of those philosophers and historians who adhered to the traditional view, and the picture increasingly claimed by sociologists to be 'realistic' on the basis of direct evidence of how scientists conduct their own business. There quickly began to appear further data from a variety of sources, including some from the 'purest' fields of scientific research, so that it could no longer be denied that something of a revolution was underway.

The development of this non-Mertonian sociology of science has depended crucially on the changing historical and philosophical perspectives outlined above, specifically, that is, upon the growth of relativism. By and large we can say that before the 1960s philosophers were content to study science as if it were nothing but logic, and historians as if it consisted solely of disembodied ideas (intellectual history). Sociologists confined themselves to its institutional structures, and only psychologists were interested in the mysterious processes of creativity itself. One of the more remarkable consequences of Popper's philosophy, despite its positivistic roots, is that its effect ultimately was to undermine the idea that science *has* any coherent logical structure. Even in his own later work (and in much of his critics) Popper noted that observations were inevitably theory-laden and themselves therefore hypothetical; there is no absolute distinction between the observational and the theoretical, and no independent framework against which to judge competing hypotheses. Thus even falsificationists can never be sure that what they have apparently falsified is actually false, any more than the logical positivists could demonstrate that what they claimed to have verified was really true. At the end of the day, some non-logical decision has to be taken as to the status of propositions and (because scientific knowledge is always *underdetermined* by the evidence) it follows that there is always more than one plausible interpretation. As the philosopher W. V. Quine pointed out, 'any statement can be held true, come what may, if we make drastic enough adjustments elsewhere in the system'.

Once scientific knowledge is accepted as underdetermined and hence inherently inconclusive, rather than as uniquely defined by the (preconceived) properties of the physical world, it becomes obvious that cultural factors of one sort or another may be among the available 'adjustments elsewhere in the system'. If even scientific knowledge-claims are partly judged by merely conventional standards of adequacy, and if (as is assuredly the case) these standards vary from time to time and place to place, then it would be no surprise to find that the knowledge itself is neither stable in meaning nor independent of circumstance. And the reason for this, of course, is that to a significant degree, scientists are able to choose between competing theories on the basis of personal or group interests and preferences that may have little to do with their actual science. Thus scientific knowledge, like other knowledge, will contain a sizeable element that is socially determined, or *relative* to the social context in which it is generated, and it is on this ground that many sociologists now entertain a fundamentally changed image of scientific knowledge-claims as simply acceptable or unacceptale rather than as true or false. Once this change is accepted, then the traditional (epistemological) distinction between scientific and social or historical knowledge is finally removed and the way opened for a full sociological analysis.

In opposition to this persuasive reappraisal, it is still argued in some circles that any 'social' study of scientific activity is capable of dealing only with the latter's peripheral, non-scientific, aspects; that genuine opportunities for a 'sociology of the conceptual and theoretical contents of science are extremely limited'. On occasions, the implication here has been that the investigator using sociological techniques is engaged in little more than sophisticated gossip-mongering. As a result, it is not surprising that the subject has generated strong feelings on both sides, such for example, that Steven Shapin, in a powerful advocacy of the new approach, can dismiss the reservations and criticisms of several distinguished scholars by saying that, for them 'it is already too late; the historical sociology of scientific knowledge has gone ahead without them'. Perhaps the opposing attitudes can be likened to Kuhn's 'incompatible modes of community life'; at least there is commonly a discernible generation gap between the warring factions. For some 'traditional' scholars, the evaluation of competing paradigms in any terms other than purely abstract notions of logical inference could not possibly have produced the coherent, widely revered, and relatively enduring body of knowledge called science. For the more 'progressive' workers, the motivation for scientists provided by particular goals, objectives and interests – as well as rational considerations – yields an immensely more plausible account of actual historical developments.

According to this new approach, paradigms are regarded by scientists not merely as assemblages of abstract theories, but rather as resources or instruments (the new model is sometimes referred to as 'instrumental') which can actually be used to *do* things. What is done then depends upon how scientists regard particular bodies of knowledge, for their attitudes, it is argued, must surely be affected by personal or social interests, such as a preference for algebra over geometry in representing the concepts of physics; the 'irrational' drive to solve certain puzzles rather than others of equal scientific interest; concerns over their group's professional image, or its opportunities for achievement; or, from the wider society, such influences as hydraulic, mechanical or computer metaphors of bodily functions.

Much work remains to be done if the new sociology of scientific knowledge is to pass from its present condition of controversy to one of established maturity. But from within the framework of epistemological relativism, scientific knowledge has special appeal as an object of study because of its traditional (and perhaps persisting) status as specially reliable, and because of its readily accessible knowledge-producing institutions. In this sense it is relatively easy to investigate. Moreover, if we can understand how scientific knowledge is produced (and if it is not markedly distinct from other kinds of knowledge) then this will be a sound basis for understanding the construction of all knowledge.

One direct approach of this sort was the 'anthropological' study of a famous neuro-endocrinology laboratory (the Salk Institute for Biological Studies in California), described by Bruno Latour and Steve Woolgar in their book *Laboratory Life* (1979). The work was undertaken with a view, as the book's subtitle has it, to revealing 'the social construction of scientific facts', in so far as it concentrates on the *process* by which scientists make sense out of their observations. The approach adopted was 'analagous with that of an intrepid explorer of the Ivory Coast', in which a perceptive non-scientist (Latour) conducted interviews and, more particularly, made close observations over a period of twenty-one months on the day-to-day activities of scientists at the laboratory bench. In this way, it proved possible to identify the principal means by which order was created out of initial disorder ('the organisation of persuasion through literary inscription', i.e. writing) but, significantly, altogether impossible to separate the 'technical or intellectual' issues from wider social concerns of the working group. Within the hallowed walls of the laboratory, nothing of a specially 'scientific' nature seemed to be going on, so that such an investigation is seen to strip science of any vestige of the cacooned mythology which it inherited from the partial accounts of philosophers.

Another approach is that associated with the so-called 'strong programme' in the sociology of scientific knowledge, currently the most influential school. According to David Bloor, one of its co-founders, the two basic principles are, first, that it is impartial with respect to truth and falsity, rationality or irrationality, success or failure and, second, that it is symmetrical in style, seeking the same types of causal explanation for scientific beliefs, no matter whether they prove ultimately to be true or false. Thus there has been a good deal of energy expended on marginal (not to say controversial) as well as mainstream science in an attempt to show that controversy is a normal part of the process of new discovery.

One of the most active themes in the strong programme has been concerned to demonstrate the 'negotiability' of knowledge. This suggestive metaphor is used to show how scientists cope (perhaps unconsciously) with the philosophical problems of the underdetermination of theories by the evidence and the theory-ladenness of observations, by exploiting interests or commitments from outside the immediate problem area as a basis for their theoretical interpretations and preferential judgements. Such interests are very often professional in nature, reflecting differences of background training or technical expertise within a given discipline, or the more widely divergent approaches of clearly defined sub-groups to an area overlapping the traditional territories of both. Two examples from the first category – on the problems of classifying plants and of replicating experiments in physics – are discussed in some detail below; an instance of the second would be the respective attitudes of nineteenth century geologists and biologists to the role of teleology and the question of what precisely was to be understood by the adaptation of an organism to its environment. In other cases particular approaches to narrow scientific problems seem clearly to draw on wider social values as, for example, in the case of an early twentieth century struggle within (even) a highly esoteric branch of mathematical statistics, which was evidently influenced by the protagonists' commitments to the then fashionable eugenics (race improvement) programme; or the arguments among French scientists of the nineteenth century over the possibility of the spontaneous generation of life, which can be seen as reflecting both moral and political attitudes to the 'external' philosophical issue of materialism. Another instance might be the contrasting interpretations of the data of modern physics mentioned in chapter 5, which tend to reflect either scientistic or broadly 'spiritual' world views, while a considerable body of recent literature has dealt with the changing historical relations of the scientific community with the church. The famous case of the influence upon Darwin of Malthus's social thought is outlined as the third example below.

At all events, it is impossible to disagree with Shapin that this new tradition in the sociology of science, controversial as it undoubtedly remains, now represents 'more than a set of theoretical and programmatic reflections on what might have been; it is also a body of practical achievements' and, indeed, an area of intense interest to contemporary scholars. The school prides itself most of all on its substantial empirical base, and it is to this that we must now turn for a fuller evaluation.

The 'Invention' of Natural Order

A good example of the part played by professional vested interests is to be found in the work of John Dean on twentieth century botanical taxonomy. It, and cases like it, are especially important because they show that while sociological explanation may be indispensable for a proper understanding of developments in science, this need not imply an 'external' model of its history. While the social factors here involved are clearly 'internal' to the science of botany, they nevertheless illustrate the considerable influence that can be exerted by sub-disciplines on the question of what constitutes botanical reality.

Traditional taxonomy has always been an observational and descriptive discipline based essentially on the pioneering work of the Swedish botanist Carolus Linnaeus (1707–78). For Linnaeus, animal and plant species were objective entities, naturally separated from each other by real and distinguishable characters. The taxonomist's task was one of discovery, for he was to identify these specific features and so expose the underlying order of the natural world. The problem of variation within the category of the species was acknowledged, but regarded as relatively unimportant because the system was designed primarily to be of practical utility, relying upon easily identified external features (in particular those of the reproductive organs) and providing a simple binary nomenclature by which a group of plants could be named 'as easily as one names a person'. The system enjoyed an enormous success after Linnaeus's own death and survived without apparent difficulty even the implication of nineteenth century evolutionary theory that species were not fixed entities after all, but only convenient abstractions in the taxonomist's mind. Indeed, it was almost that evolutionary theory and the Linnaean system strengthened each other, for the ordered hierarchy of forms was seen as proof of their natural affinities. Thus, at the beginning of the twentieth century, orthodox taxonomy had changed little since Linnaeus's day, herbarium taxonomists still seeking to describe new species on the basis of morphological characters and to provide a logical framework for the diversity of living plants.

The challenge to the traditional system came, not from professional taxonomists, but from other botanists with an ecological and experimental bent. Their research was being powerfully influenced by the rediscovery of Mendel's work on inheritance which, in its impact on Darwinian theory, encouraged great interest in population genetics and in the mechanism of speciation at the cellular level. As new techniques became available, involving the cross-breeding of varieties and species, together with the detailed analysis of intraspecific variations and chromosome behaviour, so the Linnaean system was quickly characterized by the experimentalists as a 'mouldy shroud', no longer adequate to the task in its newly complex guise. Taxonomy was at last to be relieved of its musty 'museum image', for the application of experimental methods would transform it 'from a field overgrown with personal opinion to one in which scientific proof is supreme'.

We need not examine in any detail the response of the herbarium taxonomists to such severe criticism; suffice it to say that they have typically regarded hybridization and other experimental work as peripheral to the main, and continuing, function of taxonomy, which is to provide precise descriptions and definite names. Their definition of the species still refers to populations of similar organs separated from related species by morphologically identifiable characters, while that of the experimentalists emphasizes instead the capacity for gene-exchange. In most (sexually breeding) cases, these different conceptions of the species have had little effect on the question of classification, but in the many plants where asexual reproduction occurs, the two taxonomies exist side-by-side, in 'competition'. Just such a situation occurs in the genus *Gilia* (of the phlox family) in which the *G. inconspicua* complex is divided into five distinct species by experimental taxonomists using cytological criteria, while traditional taxonomists identify only one. However, this is not simply a case of more subtle discrimination by modern methods, for with the *G. tenniflora-latiflora* complex the situation is reversed. Here the morphological taxonomists identify at least four species, while gene-exchange data persuade the experimentalists that there is just one.

This problem remains unresolved today. No consensus in favour of one system or the other has emerged, says Dean, because the alternative systems of classification are merely 'conventions designed to portray different aspects of reality and to satisfy different demands for technical prediction and control'. Thus classification is not being seen as a direct reflection of the natural world, but rather, natural reality is construed as corresponding to a classification; a classification, moreover, which defends and advances the technical vested interests of a relatively isolated group, either that of the traditional taxonomists typically working in herbaria, or of the experimentalists working in university research departments. The different interests of the two groups play a vital role in *constructing* their respective classifications, which are therefore not objective discoveries about the world, but inventions designed to emphasise particular aspects of it.

The Detection of Gravity Waves:
Replication and the Experimenter's Regress

The previous example concerned the problems posed for the integrity of an old discipline by the introduction of new techniques. This one concerns work on the elusive problem of what is to count as a competent experiment in an area of research at the very frontier of physics. It has been investigated intensively by Harry Collins.

Most physicists agree that a prediction arising from Einstein's general theory of relativity is that massive bodies moving through space will produce fluxes of gravitational radiation, often known simply as gravity waves. These might be regarded as analogous to electromagnetic radiation, and would result from catastrophic events in the universe, such as exploding supernovae or black holes. The question remains, however, as to whether traces of the vast energy generated at unimaginable distances away in space can be detected on earth as a tiny variation in the gravitational constant. This is by no means 'unreasonable' in principle, for as long ago as 1798 Henry Cavendish (in what is often referred to as 'the Cavendish experiment') succeeded in measuring the gravitational force acting between two large lead balls, even though it amounted to no more than one five-hundred millionth of their mass. Although the detection of gravity waves themselves would be very much more difficult (because they represent only a minute fluctuation *within* the other weak force), one American physicist in the 1960s thought it well worth a try. He built a detector (known as an antenna) which consisted of a massive aluminium bar suspended by a wire within a metal vacuum chamber. The chamber insulated the bar against interference from electrical, magnetic, thermal and acoustic forces, and the suspension provided seismic insulation in the shape of lead and rubber sheets. Minute changes in the length of the bar were to be detected by ultra-sensitive strain gauges glued to its surface. The signals generated were suitably amplified, freed of 'background' noise, and fed into a chart recorder; so sensitive was the whole system that the impact of light from a torch (photons) would send the recording trace completely off scale.

It is important to be clear at this stage that the search for gravity waves is a perfectly respectable activity within physics and that a successful outcome would be considered of the greatest importance. There is widespread agreement both about the type of apparatus needed and the physical theory which underlies its use. The problem, unfortunately, is that of achieving the essential agreement on what would constitute a successful outcome. In the early 1970s, the original investigator produced results that were thought by many physicists to be, as it were, *too* good; the amount of radiation that he claimed to have detected seemed far too great to conform with contemporary cosmological theory. If his interpretation of measurements was correct, argued his critics, the universe would soon be completely 'burnt out'.

Nevertheless, the original work was significantly refined in response to initial criticism and it encouraged teams in other laboratories to build their own antennae, in most cases with the intention of repeating the observations and refuting the original claims. But it was just such attempts that exposed the difficulties implied by replication in a pioneering area of research where there were quickly developing disagreements over the meaning of observations, the reliability of experiments, and so on. Collins identifies one particular overall problem and designates it the paradox of the experimenter's regress. In essence this states that any experimental check on the results of an earlier experiment can never be clear-cut, simply because the act of experimentation is itself a matter of skilful practice. Thus a third experimental test would be necessary to check on the quality of the second, and so on, ad infinitum. In the context of the search for gravity waves, a successful outcome would naturally depend on whether such fluctuations really do reach the earth and whether they are detectable. To check on this a suitable antenna would have to be built and used. But the suitability of the antenna could be known only after it had produced a successful outcome. And there would be no way of knowing what a successful outcome is until . . . etc. Thus the experimenter's regress can be broken only by finding some test of the quality of an experiment which is unaffected by the outcome of the experiment itself. In the case of gravity waves there was no such independent criterion.

Almost all of the antennae built to check the original knowledge-claims concerning gravity waves were (and are) of very similar design to the first, although often of enormously greater sensitivity. What happened when their results began to come in was, briefly, that the (few) physicists committed to the reality and detectability of the fluctuations, judged as competent those experiments that appeared to detect them; experiments that failed to do so were correspondingly 'incompetent'. On the other hand, (the many) scientists sceptical of the original claims took the opposite view: for them a positive report of gravity waves simply indicated an experiment incompetently performed or interpreted (e.g. the fluctuations on the chart were attributable to sources other than large bodies in space). Conflicting views about the natural world were thus being used, so to speak, to calibrate the actual experiments.

The situation continued to be one of great complexity, however, because in almost every case, experimental data and arguments advanced in opposition to the original claims were themselves widely criticized. Explanations for the many discrepancies took many forms; most of them were technical criticisms of design procedures or statistical analysis, but some referred to personal qualities of the scientists, and even to the necessity to invoke psychic forces to explain certain results. Although the background 'cherished beliefs' of the orthodox community of physicists were for a time in turmoil, by 1975 the original claims (though not the possibility that gravity waves may be detectable on earth) were universally regarded as false. What had happened here, argues Collins, was that the participants had engaged in 'negotiations about the meaning of a competent experiment', had reached consensus about their problematic empirical phenomena, and established for themselves a distinct scientific sub-culture which safeguarded their shared background assumptions. Thus the demise of the original claims 'was a social . . . process' having little to do with the objective, universal criteria of any idealized scientific method.

As a post-script to this case, it is worth mentioning that an attempt was made by the sceptics to break out of the experimenter's regress by instituting a 'test of a test' in the shape of the physical calibration of the competing antennae. The method used involved the injection of pulses of electrostatic energy into the aluminium bar, on the assumption that a 'suitable' antenna would surely detect pulses the existence of which no one doubted. By this means, the various competing systems (which detected no gravity waves) were shown to be at least as sensitive as the original, but of course this test begged the question of the suitability of the energy used for calibration purposes (i.e. it assumed that electrostatic energy *would* be suitable). Whereas in 'normal' science this would seldom be a problem, in the 'extraordinary' circumstance where the real nature of the phenomenon under investigation is not known, it is another matter entirely. In the event, and despite the objections of the individual advancing the original claims whose freedom to interpret his results was being severely curtailed, the orthodox community decided that the calibration signal was 'reasonable' and that the time had come to draw the episode to a close.

Had the argument gone the other way – that is, if 'gravity waves' had been *defined* as something detectable by the original antenna but not by the other, unsuitable, ones, and *explained* in terms of their difference – it would have raised radical possibilities for physics which consensus showed the community was not prepared to entertain.

Cultural Resources of Evolutionary Theory

In traditional accounts of Darwinian theory historians have often been anxious not simply to describe or explain social influences, but rather to explain them away. This seemed an attractive strategy because the acknowledged ideological influences upon other evolutionists, for example Lamarck, were thought to have led him into inaccurate scientific speculation. Darwin, it was said, was influenced only by objective evidence from the natural world, and it was this that made it possible for him to explain the actual mechanism that accounts for evolutionary change.

By contrast with this, contemporary historical sociology tends to see wider cultural resources as the inevitable background against which all kinds of thought, moral, political, and scientific must develop. In the case of Darwin (and of Alfred Russell Wallace, who independently reached the same conclusion) there is now an enormous literature intended to clarify the position with respect to two particular social influences, that of the commercial animal and plant breeders (or 'fanciers'), and that of Thomas Malthus's famous *Essay on Population* (1798). It will be the task of this final example to provide a brief summary of the main points.

An original research method undoubtedly developed by Darwin himself was that of recording over long periods of time the small variations seen in animals and plants under domestication. By this means he intended to provide the necessary support for his belief that biological structures do change and that these changes can be inherited. For the purpose of building up his detailed records, Darwin was obliged frequently to share the company of commercial breeders whose motivation was simply profit through successful business, certainly not knowledge for its own sake. Such intercourse was justified by his assumption that selection of morphological features by the natural environment was analogous to that in the artificial (where it simply satisfied market demands); that is, both were functionally 'adaptive' in essentially the same sense. Darwin even went so far as to suggest that there must be an equivalent agent, 'a being infinitely more sagacious than man' who performed the natural selection for the 'improvement of each organic being in relation to its . . . conditions of life', just as the breeders did for the purposes of commerce. Although he professed to regard 'natural selection' as only a metaphor for the impersonal laws of nature, it appears that it may have had convenience in allowing him to side-step the problem of showing that natural and artificial selection were *in fact* equivalent. This he could not do simply for lack of sufficient evidence from the natural world.

From what source, then, did Darwin derive his idea for the explanatory mechanism that he himself regarded as the one original feature of his work? Both he and Wallace, it now seems certain, experienced a dramatic insight which allowed them to generalize the well-known 'laws' governing human population, set out so persuasively by Malthus, to the much wider arena of animal and plant life. The evidence so meticulously collected on intraspecific variation, both artificial and natural (remember that Darwin had travelled extensively as a young man, collecting specimens from many parts of the world), could now be interpreted in a new light. If population tended to increase in geometrical ratio, but food supplies only arithmetically, the growth in numbers would surely be kept in check, but the poor and weak would suffer proportionately more than the strong and better off. This law would also operate in favour of the 'fittest' biological varieties, and it therefore provided a ready-made general account of evolutionary change. That it remained still an abstract account was unavoidable, because at the time there was no knowledge of the genetic mechanism by which favourable characters may be inherited under selective pressure.

It appears not only that Darwin was heavily indebted for the form and content of his scientific theory to the language and attitude of the commercial breeders and to the arguments and predictions of Malthus, but that these factors also aided the theory's rapid acceptance by the scientific world and the wider society alike. On the one hand, Darwin's reference to the 'wiser being' ultimately responsible for the very laws of nature which promoted beneficial adaptation, made reconciliation with traditional theology that much easier and, on the other, the doctrine of the survival of the fittest rang poignantly true for a Victorian society which largely accepted inequality and competition as necessary features of communal life, and was itself having to adjust rapidly to the stresses of industrialization. What, at very least, seems *not* to have been the case – and this is of greatest significance in the context of the new sociological explanation – is that Darwinian theory was based directly upon observations of facts about the natural world.

Perhaps the most useful general conclusion to be drawn from this example is not so much that the intellectual content of science is commonly affected by 'external' cultural resources, but that there tends to be a common background of 'philosophical' debate which is routinely deployed in social and political argument as well as in natural science itself.

Epilogue

It may be that the new sociology of scientific knowledge is not so remarkable after all. Perhaps it is no great surprise to learn that scientists are much like other people, even when actually doing their science. Did we ever really believe that they hung up their individual personalities while putting on their communal white coats? Of course we are all well aware that individuals can be wrong, but then so too can communities; the important point surely is that the social processes of negotiating knowledge, far from exposing the frailty and fallibility of the scientific community, are more likely to minimise them. It is not the revelation that scientists fail to produce 'true' knowledge that should impress us but, rather, that despite being 'only human', they produce knowledge that remains the most reliable we *can* have.

Throughout the text above the words 'knowledge' and 'belief' have been used almost as if they were interchangeable. This may be acceptable for most purposes, but it is not satisfactory in the stricter philosophical sense. The discipline that we have referred to as the sociology of knowledge would perhaps better be named the sociology of belief, since it has relatively little to do with any grounds there may be for accepting particular beliefs as genuine knowledge. One of the problems for this discipline in the context of science is that many people (especially philosophers) have always tended to feel that the effect of there being a sociological explanation for a belief is automatically to discredit it. It seems to have been for this same reason that many sociological studies have analysed religious beliefs or alien ideologies and, by the same token, have ignored science. With the rise of relativism, and perhaps with the growth of the anti-science movements, it has not only become permissible to analyse scientific beliefs sociologically, but also fashionable to discredit them. Thus the pendulum has swung all the way from the myopic obsession of philosophers with the rational justification of scientific knowledge, to the single-minded pursuit by sociologists of the social construction of scientific beliefs.

Not that extreme views are a bad thing. They typically encourage the majority of individuals to find shelter somewhere in between, in a position that may be less exhilarating (in the way that sophisticated academic games can be) but is often nearer to the truth. Of course, a formal objection to any notion of the universal social conditioning of knowledge would be that it is self-contradictory; there can be no reason to accept the claims of a sociology of knowledge which must themselves be so conditioned. As was said at the beginning of this chapter, we therefore seem to need a sociology of the sociology of science to tell us something of what motivates the sociologists.

This objection aside – and perhaps it has no more practical importance that the familiar critique of induction – it remains true that sociological investigations have not undermined the essential foundations upon which the scientific enterprise is built, namely widespread consistency and reliability. Only someone already committed to the 'strong programme' will fail to see that it can be as doctrinaire and authoritarian in its own way as the traditionalist views that it assails. The insight that scientists are guided by whatever promotes their self-interest may be the truth, but it is no more the whole truth than the idea that they respond only to rational analysis. A partial view which attacks 'objectivity' because it is actually no more than 'intersubjective consensus'; which construes extraordinary science almost as if it were normal science; or which blurs the very real distinction between the rarified world of philosophical debate and the pragmatic world of day-to-day science, may relate something of considerable importance about the processes of scientific discovery and growth at the expense of undervaluing the continuing importance of the scientific tradition. Although we now see that it is foolish to make claims for the absolute truth of scientific knowledge, this is far from saying that it is nothing but superstition. Science, both as a body of knowledge and as a method for acquiring it, will no doubt always give us an imperfect model of the world, but it is still likely to remain the best model that we have.

AUTHORITY AND FRAUD

Stanley Milgram's classic work *Obedience to Authority*, discussed in chapter 7, provided disturbing evidence of the passive acceptance that lay people can display towards the commands of an imperious figure representing an apparently all-knowing 'science'. It reminds us of the dangers inherent in absolute authority and of the urgent need for scientists to divest themselves of it. The very notion of obedience to a powerful elite seems contrary to the ideals not only of science, but also of modern democratic society, which encourages individual freedom of thought and action. If the ideals of democracy are essentially those also of science, so that each individual is free to make his own observations and to use his own rational faculties in pursuing the goal of intersubjective consensus, coercion by authority is simply not necessary. Science, we are told, is public knowledge freely available to all.

Of course, it is obvious that the practitioners and institutions of science continue to wield enormous authority, just as they have for a hundred years as the increasingly successful discoverers and repositories of specialized knowledge. As teachers, industrial administrators, advisors to governments, expert witnesses, technical (including military) innovators, even as all-round gurus uniquely fitted to pronounce on such great issues as the Ascent of Man or the Wisdom of the West, scientists are presented, or present themselves, as embodiments of an authority that can be safely trusted and followed. As qualified professionals representing what is widely regarded as the most reliable form of modern knowledge, they are still treated with exceptional respect, much as the theologians and clergymen of the early nineteenth century were respected as representatives of their then-dominant world view. Such authority, and the respect that it engenders, seems to be an inevitable consequence of, if not a pre-requisite for, the highly specialized society that is organized around a complex division of (intellectual) labour. In practice, it is quite impossible for the individual citizen or the individual scientist to check on the accuracy of everything that others say, or to gain all the know-how necessary to conduct his business; the price of access to specialized knowledge must therefore be trust in the authority of experts.

But whom are we to trust when experts disagree? More and more frequently we are becoming accustomed to the spectacle of scientists in conflict with each other. Specialized knowledge of a highly technical kind is advanced in advocacy of a particular point of view – typically, for example, in the context of the nuclear power debate (see chapter 7) – only to be challenged by the advocates of another. The conflict may occasionally centre on the question of what the facts really are, but more commonly on different interpretations of their social significance. What is perhaps most striking is that expert opinion is increasingly identified as just that, *opinion*, and often by lay individuals *despite* their claim to no technical knowledge other than that of the well-informed citizen. This suggests a declining respect for the authority of scientific experts and, concurrently, a bolder willingness to 'take them on'. As expert opinion is seen to be necessarily partial (i.e. technical considerations alone are seldom a sufficient basis for major policy decisions), and as the technique of arguing persuasively against it is better developed, we may expect a continuation of the trend. Perhaps the roots of this new confidence are to be found in the warning note sounded by the Milgram experiment, in the anti-science movements of the 1960s, and particularly in the new sociology of scientific knowledge outlined above. If science is just as likely to be influenced by social forces as other forms of human enquiry, what basis could there be for its traditional claim to special respect? If there is none, then perhaps Feyerabend is right: science (which of course, is acknowledged by everyone to have an indispensable role in the limited sphere of technicalities) must compete in the struggle for an overall world-view with astrology, mysticism and magic. . . .

To make matters worse for the traditional view of science as *the* (benign) cognitive authority, there is growing evidence that scientists are just like other people, in their worst moments as well as in their better. In the light of what was said earlier in the chapter, this should not be a revelation of very great impact. Nevertheless, it cannot be denied that the documentation of cases of actual fraud, both from past and present-day science, represents yet another thorn in the side of the conventional ideology. In a recent book with the somewhat sensational title *Betrayers of the Truth* (1982), William Broad and Nicholas Wade set out to investigate science specifically through its 'pathology'. Their basic contention – the same as that of the new sociology – is that the traditional view derives from the work of historians, philosophers and orthodox sociologists, who have studied science not for its own sake but rather from the perspectives of their own disciplines. Cases of fraud, however, 'provide telling evidence not just about how well the checking system of science works in practice, but also about the relations of fact to theory [and] about the motives and attitudes of scientists'. It is not enough to reject their analysis simply because it is conducted on a 'popular' level by journalists rather than on an academic level by scholars; their insight on the self-deception, gullibility, greed, and outright dishonesty of some practitioners of science is painfully relevant, for it tells us a good deal that is important about the scientist's twin goals of discovery and recognition.

The main point made by Broad and Wade is that, more often than we suspect, these apparently complementary goals can be in conflict. While everyone accepts that the scientist's role is to discover new knowledge about the world and that he has an absolute right to appropriate reward for priority, it can happen, especially under the pressure of modern careerism, that the drive for recognition can distort the normal processes of discovery. In short, the ambitious or anxious scientist whose experimental data are not exactly as expected, or have failed to gain general acceptance, faces many temptations to 'doctor' them artificially, or perhaps to invent better ones. It seems that even the most eminent and successful scientists are not exempt, for there is strong (and scholarly) evidence that Ptolemy made false claims about astronomical measurements, and that Galileo, Newton and Mendel all exaggerated the importance of their experimental results, or 'improved' their calculations, in order to confirm preconceived theories and enhance their personal reputations.

The sorts of temptations to which these men succumbed increased considerably as science quickly became professionalized in the present century and individuals found themselves in competition for positions of influence and authority, with their attendant financial rewards. The evidence to date suggests that those most susceptible in modern science tend to be at the extremes of the research enterprise, either working in isolation or in one of the huge and impersonal 'research mills'. (Results-orientated biologists seem to be at greater risk than scientists in the 'harder', more mathematical, and more securely theoretical, disciplines.)

A famous case in the former category is that of the British psychologist Sir Cyril Burt. At the time of his death in 1971, Burt's long record of work on the inheritance of human intelligence was acclaimed as of fundamental importance; he himself was widely regarded as the greatest educational psychologist of his day. Yet within three years, as a result of fortuitous investigations by a psychologist who had never before worked in Burt's field, the great man's reputation was in shreds. For thirty years, it seems that Burt had been so determined to 'prove' his theory that the genetic inheritance of intelligence is some three times as important as environmental influence, that he had resorted not only to the armchair invention of data (principally on the IQs of separated identical twins), but even to the invention of co-workers, who published papers in the psychological literature giving strong support to the master's own views! So persuasive was his rhetoric, so impressive his statistical analysis, and so conclusive his 'evidence' from 'on-going research', that the authoritative and urbane professor was never seriously challenged. Only after his death were many of the data shown to have been fabricated.

Even Burt, it appears, had neglected to make suitable alterations in the correlation coefficients between the IQ scores of his separated twins from which, in a number of studies, he had evidently 'worked back' (in time-honoured schoolboy fashion) to his alleged original samples. Not only were these suspiciously large (twins are very seldom separated) and ever-growing, but different samples would assuredly have yielded different coefficients had they been genuine.

A point made by Broad and Wade is that the frailties of Galileo, Newton and Mendel have been essentially 'covered-up' by the scientific community because the theories for which they 'lied' proved ultimately to be correct. Burt, by contrast, had the misfortune, as they put it, that he 'lied for truth and was wrong' (for the environmental component in IQ is now considered to be much more important than his own doctrine allowed). At any rate, he has not been forgiven, and nor have the several individuals who in recent years have been exposed as frauds, even when the bold theories for which they manufactured false data are still regarded as perfectly plausible. (A strong censorious reaction is, of course, reassuring for the public image of science; where fraud cannot be covered-up, the cheats responsible must be seen to be humiliated and eclipsed in order to sustain the 'official' view that their actions are so exceptional as to be altogether insignificant for science as a whole.) Almost all of these cases have concerned young scientists working in cancer research, immunology, or some other frontier field of the huge and multi-faceted biomedical research enterprise, characterized by Broad and Wade as the 'paper factories of the lab chiefs'. There is an important sense in which the modern phenomenon of the hierarchically structured research machine under the leadership of a single eminent individual could actually promote careerism, or possibly corruption, for it is a widely accepted tradition of scientific publication that the name of the laboratory's head appears in the author line of papers describing work done by his junior colleagues. The leader thus accumulates an ever-increasing reputation, at least in part at the expense of others; moreover, he claims credit for everything that goes well, but at once tends to deny responsibility when fraud is discovered. (This is another aspect of the problem of accountability discussed in chapter 8.) But because the publication of papers is the only way to a scientific reputation, the younger workers are obliged to accept the situation as it is, no doubt in the hope that, in the fullness of time, they too will reach the top. It is not surprising that cynicism can develop as the competitive pressure to amass credit for numbers of papers gains the upper hand over the dispassionate pursuit of truth for its own sake. For some, the result can be a fall from grace: cutting corners, (sometimes unconsciously) ignoring unfavourable results and improving good ones, and perhaps even calculated deception.

That this happens obviously tells us something important about scientific behaviour under competitive pressure. But fraud also tells us something about scientific methodology itself. According to traditional wisdom, the objectivity of science is guaranteed by the peer review system and the replication of experiments. But there is further disturbing evidence that the former can operate as little more than an incestuous old-boy network in which proposals for support are conveniently reviewed by trusted colleagues from within the applicant's own elite group; so far as 'internal' criteria are concerned (see pages 142–3) this is inevitable, for they would be the only scientists capable of informed evaluation. This same mechanism of mutual back-scratching can operate in the appointment of external examiners for higher degree candidates, whilst not everyone is satisfied with a refereeing system in which the authors of submitted papers are known, but the referees themselves anonymous. Clearly names on a paper, or even its laboratory of origin, are likely to have some influence in determining the perceived value of its contribution to knowledge, such that some critics would argue that scientific eminence is more a matter of gamesmanship than of merit.

As to the replication of experiments, we have already seen some of the problems in the case study above on gravity waves. Although replication is traditionally quoted as the unique self-correcting mechanism of science (and techniques employed are supposed to be described in the 'Methods' section of a paper in such a way that it is genuinely feasible), in practice it is very much the exception, not the rule, simply because there is no reputation to be made out of repeating someone else's work. What is regarded as a suitable validation procedure for knowledge in modern empirical science is the direct pragmatic question: Does it work? Broad and Wade suggest an 'invisible boot' operating in a way somewhat analogous to Adam Smith's 'invisible hand' (see page 142). The invisible boot 'kicks out all the incorrect, useless, or redundant data in science . . . [and] over time it stamps out the nonrational elements of the scientific process, all the human passions and prejudices that shaped the original findings, and leaves only a desiccated residue of knowledge, so distant from its human originators that it at last acquires the substance of objectivity'.

Unfortunately, it is impossible to know how widespread fraud may be, in part because the excessive proliferation of scientific papers ensures that many go unread, unquoted, and certainly unreplicated. However, all papers contribute to the convenient myth of objectivity by employing an impersonal style enforced by current editorial convention. In this sense, the research paper is itself a fraud which bolsters the authority of science by the illusion of reflecting a universal scientific method. A paper that more accurately represented the nature of scientific thought might begin, rather than end, with a discussion of the possible significance of the hypothesis tested, and be followed by a description of the new data and the techniques used to collect them. To present 'results' in cold detachment from their scientific context – as authors are required to do – is to cling to the long-discredited notion of the mind as a blank receptacle for data from the external world, and to the idea of scientific method as unadulterated inductivism.

The exposure of fraud contributes to recent work in sociology by emphasizing the fallible, more human, elements of scientific behaviour. The logical structure of scientific testing may be its cardinal characteristic, but this alone cannot define what science is, nor can it justify the separation of science into a category altogether different from that of other creative and intellectual activities. Earlier in the book (chapter 4) we examined some traditional philosophical solutions to the problem of the demarcation of science from non- or pseudo-science; it now remains to say a few words about what boundaries, if any, may still be discernible from the new sociological point of view.

PROTOSCIENCE, PARASCIENCE AND PSEUDOSCIENCE

It must be obvious that the problem of demarcation is now a great deal more complex than it was a generation ago. At least, this must be so if we are persuaded that science is influenced by social forces in much the same way as other areas of organized knowledge. But while the traditional criteria of demarcation may now seem inadequate, we still tend to act as if we know intuitively what science and non-science really are. For instance, few people would hesitate in making the distinction between the following pairs of subjects: astronomy and astrology; chemistry and alchemy; neurology and phrenology, even though closer examination might convince most of them that substantial areas of awkward grey exist between the clearly black and white.

Although it would be convenient to concentrate solely upon the distinction between science and non-science, as tended to be the pattern in the past, the title of this section indicates that the problem is now bedevilled by the use of several other terms, as if they each had precise and agreed meanings. An important point to make here is that the term 'non-science' is often used in a relatively neutral sense to describe the study of literature, metaphysics or theology, subjects which do not claim to be scientific and are not defensive about it. The other terms, however, are more clearly normative, even though they still mostly refer to kinds of activities that fall within the category of non-science. Thus 'pseudoscience' clearly implies the burden of the false pretender; we generally understand it to refer to an activity and a set of beliefs that claim to be genuinely scientific, but are not recognized as such by the scientific establishment. The term 'parascience' is more complex for it is used to describe quite a wide spectrum of activities. For example, the 'paramedical' science are simply those ancillary to clinical practice; in this particular case they are recognized to be genuinely scientific and indispensable as the basis of modern medicine, yet they are still somehow subservient to the great profession itself (which, ironically, could be said to incorporate many elements that are clearly non-scientific). At the other extreme, 'paranormal' phenomena in general, such as clairvoyance and telepathy, fall within the the highly controversial area of 'parapsychology' – dismissed by the majority of scientists as simply pseudoscience, yet gaining the guarded support of a minority. Finally, the notion of 'protoscience' seems unambiguously to refer to disciplines with the potential for genuine scientific status, but which for the present are in a state only of 'becoming'.

In all this semantic wrangling, we have been careful to avoid confronting the one really important question of what exactly it is that we mean by science itself. Yet without a satisfactory answer the other distinctions would, of course, be meaningless. This is really the crux of the dilemma, for the impact of the new sociology of scientific knowledge has been to make any universal definition seem almost impossibly elusive. Even though the sciences of astronomy and chemistry undoubtedly evolved in part from astrology and alchemy, it should nevertheless surprise no-one that the exact moments of metamorphosis or maturity cannot be identified, any more than the precise point of transition between the modern disciplines of physics and chemistry. If we become too obsessed with the design of the straightjacket within which we intend to confine nature, we run the risk of forgetting that nature resolutely refuses to be thus confined.

We have, in effect, returned to the main point of this chapter: the traditional account of science, which was given considerable weight in earlier chapters, is now under constant bombardment from the new generation of radical philosophers and sociologists. But a clear consensus in favour of one view or the other seems still to be far in the future. To argue that formal demarcation criteria for science *can* be set down is to acknowledge the continuing credibility of the orthodox view; this is a position that is still stoutly defended. On the other hand, to argue that demarcation is determined merely by social and historical factors is to accede to the contemporary trend; this is the position from which the most damaging attacks are now mounted. Thus it was not the intention of the chapter to leave the reader overwhelmed by the rising tide of relativism, but rather to counterbalance the massive stronghold of the positivists which, for many, remains bruised but unbroken. Nor, indeed, was it the intention to imply that only two possible positions can be taken in the debate. Any such conclusion would surely be a gross over-simplification. That the main conflict should be conducted between the extremes is entirely to be expected, but this does not entitle us to assume that the one or the other must hold a monopoly of the 'truth'. Should a consensus ultimately emerge as to which is the more accurate portrait of science, it would be surprising if the victor were to go uninfluenced by the vanquished. At any rate, for the reader whose interest is sufficiently aroused, it would be a relatively easy matter to pursue the debate further. In the meantime, continuation of so healthy a struggle can do nothing but good for our attempt to understand science better.

Bibliography

The following is a list of the most important books and articles which have been consulted. Dates given refer to the actual volumes used and are therefore not necessarily those (usually of first editions) which appear in the text. With a small number of exceptions, duplication has been avoided by citing a reference in connection only with the first chapter to which it applies.

Introduction

Hudson, L. *Contrary Imaginations* (London, Methuen, 1966).
   Report of the Committee on higher Education appointed by the Prime Minister ('Robbins Report'), *Cmnd 2154* 1963.
Snow, C. P. *The Two Cultures and A Second Look* (New York, Mentor, 1964).

Chapter 1   The Structure of Science

Campbell, N. *What is Science?* (New York, Dover, 1953).
Russell, B. *The Problems of Philosophy* (Oxford, University Press, 1964).
Ziman, J. *Reliable Knowledge* (Cambridge, University Press, 1978).

Chapter 2   Scientific Argument: The Role of Logic

Bradbury, F. R. (ed.) *Words and Numbers* (Edinburgh, University Press, 1969).
Cohen, M. R. & Nagel, E. *An Introduction to Logic* (London, Routledge & Kegan Paul, 1963).
Pauling, L. *Vitamin C and the Common Cold* (London, Ballantine, 1972).
Salmon, W. C. *Logic* (Englewood Cliffs, NJ, Prentice-Hall, 1973).

Chapter 3   The Scientific Attitude: Science in Practice

Durkheim, E. *Suicide: A Study in Sociology* (1897; London, Routledge & Kegan Paul, 1952).

Hales, S. *Vegetable Staticks* (1727; London, MacDonald, 1969).
Harvey, W. *The Circulation of the Blood* (London, Dent/Everyman, 1963).
Wells, W. *An Essay on Dew and Several Appearances connected with it* (London, Taylor & Hessey, 1814).

Chapter 4   Philosophies of Scientific Method: Theories of Science

Burniston Brown, G. *Science: Its Method and its Philosophy* (London, Allen & Unwin, 1950).
Chalmers, A. F. *What is this Thing called Science?* (Milton Keynes, Open University, 1982).
Feyerabend, P. K. *Against Method: Outline of an Anarchistic Theory of Knowledge* (London, New Left Books, 1975).
Grazia, A. de (ed.) *The Velikovsky Affair* (London, Sidgwick & Jackson, 1966).
Harré, R. *The Philosophies of Science* (Oxford, University Press, 1978).
Hempel, C. G. *Philosophy of Natural Science* (Englewood Cliffs, NJ, Prentice-Hall, 1966).
Kuhn, T. S. *The Structure of Scientific Revolutions* (Chicago, University Press, 1970).
Lakatos, I. & Musgrave, A. (eds) *Criticism and the Growth of Knowledge* (Cambridge, University Press, 1970).
Losee, J. *A Historical Introduction to the Philosophy of Science* (Oxford, University Press, 1977).
Magee, B. *Popper* (London, Fontana, 1973).
Medawar, P. B. *Induction and Intuition in Scientific Thought* (London, Methuen, 1969).
Medawar, P. B. *The Art of the Soluble* (London, Methuen, 1967).
Mill, J. S. *System of Logic* (London, Longmans, 1891).
Popper, K. R. *The Logic of Scientific Discovery* (London, Hutchinson, 1977).
Popper, K. R. *Conjectures and Refutations* (London, Routledge & Kegan Paul, 1972).
Popper, K. R. *Objective Knowledge* (Oxford, University Press, 1973).
Velikovsky, I. *Worlds in Collision* (London, Abacus, 1978).
Whewell, W. *The Philosophy of the Inductive Sciences,* vol. II (London, Parker, 1840).

Chapter 5   The Nature of Science: Physical, Biological and Social Sciences

Capra, F. *The Tao of Physics* (London, Fontana, 1979).
Eddington, A. *The Nature of the Physical World* (Cambridge, University Press, 1929).
Forman, P. 'Weimar Culture, Causality and Quantum Theory' *Historical Studies in the Physical Sciences* 3, 1971, pp. 1–115.
Harré, R. (ed.) *Scientific Thought, 1900–1960* (Oxford, University Press, 1969).

Heisenberg, W. *The Physicist's Conception of Nature* (London, Hutchinson, 1958).

Hull, D. *Philosophy of Biological Science* (Englewood Cliffs, NJ, Prentice-Hall, 1974).

Nagel, E. *The Structure of Science* (London, Routledge & Kegan Paul, 1979).

Rudner, R. S. *Philosophy of Social Science* (Englewood Cliffs, NJ, Prentice-Hall, 1966).

Talbot, M. *Mysticism and the New Physics* (London, Routledge & Kegan Paul, 1981).

Teilhard de Chardin, P. *The Phenomenon of Man* (London, Collins, 1965).

Waddington, C. H. *The Scientific Attitude* (London, Penguin, 1948).

Whitehead, A. N. *Science and the Modern World* (Cambridge, University Press, 1933).

Winch, P. *The Idea of a Social Science and its Relations to Philosophy* (London, Routledge & Kegan Paul, 1977).

Chapter 6 Social Studies of Science and Technology

Averch, H. A. *A Strategic Analysis of Science and Technology Policy* (Baltimore and London, Johns Hopkins University Press, 1985).

Barnes, B. (ed.) *Sociology of Science* (London, Penguin, 1972).

Ben-David, J. 'Scientific Entrepreneurship and Utilization of Research' in Barnes, B. *Sociology of Science*, pp. 181–7.

Bernal, J. D. *The Social Function of Science* (London, Routledge, 1944).

Blackett, P. M. S. 'Memorandum to the Select Committee on Science and Technology' *Nature* 219, 1968, pp. 1107–10.

Blume, S. S. (ed.) *Perspectives in the Sociology of Science* (Chichester, Wiley, 1977).

Branscomb, L. 'Industry Evaluation of Research Quality: Excerpts from a Seminar' in La Follette, M. C., *Quality in Science* (Cambridge, Mass., and London, MIT Press, 1982).

Braun, E., Collingridge, D. & Hinton, K. *Assessment of Technological Decisions – Case Studies* (London, Butterworths, 1979).

Cardwell, D. S. L. *Technology, Science and History* (London, Heinemann, 1972).

Carter, C. F. & Williams, B. R. *Industry and Technical Progress* (Oxford, University Press, 1957).

Central Advisory Committee for Science and Technology. Technological Innovation in Britain (London, Her Majesty's Stationery Office, 1968).

Crane, D. *Invisible Colleges: Diffusion of Knowledge in Scientific Communities* (Chicago, University Press, 1972).

Cotsgrove, S. F. & Box, S. *Science, Industry and Society* (London, Allen & Unwin, 1970).

Ellis, N. D. 'The Occupation of Science' *Technology and Society* 5, 1969, pp. 33–41. Reprinted in Barnes, B. *Sociology of Science*, pp. 188–205.

Freeman, C. *The Economics of Industrial Innovation* (London, Penguin, 1974).

Galbraith, J. K. *The New Industrial State* (London, Penguin, 1966).

Goldsmith, M. & Mackay, A. *The Science of Science* (London, Penguin, 1966).

Green, K. & Morphet, C. *Research and Technology as Economic Activities* (London, Butterworths, 1977).

Greenberg, D. S. *The Politics of American Science* (London, Penguin, 1969).

Hagstrom, W. O. *The Scientific Community* (New York, Basic Books, 1965).

Hobsbawm, E. J. *Industry and Empire* (London, Penguin, 1969).

Hollomon, J. H. in Trybout R. A. (ed.) *Economics of Research and Development* (Columbus, Ohio, University Press, 1965), p. 253.

Johnson, P. S. *The Economics of Invention and Innovation* (London, Robertson, 1975).

Langrish, J., Gibbons, M., Evans, W. G. & Jevons, F. R. *Wealth from Knowledge* (London, Macmillan, 1972).

Merton, R. K. *Science, Technology and Society in Seventeenth Century England* (New York, Howard Fertig, 1970).

Merton, R. K. 'The Institutional Imperatives of Science' in Barnes, B. *Sociology of Science*, pp. 65–79.

Mulkay, M. J. *The Social Process of Innovation* (London, Macmillan, 1972).

Mulkay, M. J. 'Cultural Growth in Science' in Barnes, B. *Sociology of Science*, pp. 126–41.

Mulkay, M. J. *Science and the Sociology of Knowledge* (London, Allen & Unwin, 1979).

Pevitt, K. & Warboys, M. *Science, Technology and the Modern Industrial State* (London, Butterworths, 1977).

Price, D. de Solla *Little Science, Big Science* (New York, Columbia, 1963).

Price, D. de Solla 'Science and Technology: Distinctions and Interrelationships' in Barnes, B. *Sociology of Science*, pp. 166–80.

Rose, H. & Rose, S. *Science and Society* (London, Penguin, 1960).

Spiegel-Rosing, I. & Price, D. de Solla *Science, Technology and Society* (London, Sage, 1977).

Williams, B. R. *Technology, Investment and Growth* (London, Chapman & Hall, 1967).

Williams, B. R. (ed.) *Science and Technology in Economic Growth* (London, Macmillan, 1973).

Chapter 7   *Ethical Dimensions of Science*

Ashby, E. *Reconciling Man with the Environment* (London, Oxford University Press, 1978).

Bernal, J. D. *The Social Function of Science* (London, Routledge, 1944).

Black, M. 'Is Scientific Neutrality a Myth?' in Lipscombe, J. & Williams, B. *Are Science and Technology Neutral?* pp. 40–53.

Braun, E. & Collingridge, D. *Technology and Survival* (London, Butterworths, 1977).

Bunyard, P. & Morgan-Grenville, G. *Nuclear Power* (Godalming, Surrey, Ecoropa, 1980).

Cameron, I. & Edge, D. *Scientific Images and their Social Uses* (London, Butterworths, 1979).

Corson, B. B. & Stoughton, R. W. 'Reactions to Alpha, Beta-Unsaturated Dinitriles' *Journal of the American Chemical Society* 50(2), 1928, pp. 2825–37.

Diener, E. & Crandall, R. *Ethics in Social and Behavioural Research* (Chicago, University Press, 1978).

Fieser, L. F. *The Scientific Method* (New York, Reinhold, 1964).

Flew, A. G. N. *Evolutionary Ethics* (London, Macmillan, 1967).

Gowing, M. & Arnold, L. *The Atom Bomb* (London, Butterworths, 1979).

Humphreys, L. *Tearoom Trade* (Chicago, Aldine, 1975).

Huxley, T. H. & Huxley, J. S. *Evolution and Ethics* (London, Pilot, 1947).

Illich, I. *Medical Nemesis: The Expropriation of Health* (London, Marian Boyars, 1976).

Kahn, H. & Weiner, A. *The Year 2000 – A Framework for the Next Thirty-Five Years* (New York, Macmillan, 1967).

Lewis, L. C. (ed.) *Report from Iron Mountain on the Possibility and Desirability of Peace* (London, Penguin, 1968).

Lipscombe, J. & Williams, B. *Are Science and Technology Neutral?* (London, Butterworths, 1979).

May, W. F. 'Code, Covenant, Contract or Philanthropy' *Hastings Center Report* (New York) 5 (6), 1975, pp. 29–38.

Meadows, D. H., Meadows, D. L., Randers, J. & Behrens, W. W. *The Limits to Growth* (London, Pan, 1974).

Milgram, S. *Obedience to Authority* (London, Tavistock, 1974).

Monod, J. *Chance and Necessity: an Essay on the Natural Philosophy of Modern Biology* (London, Collins, 1972).

Polanyi, M. *The Logic of Liberty* (Chicago, University Press, 1958).

Polanyi, M. 'The Republic of Science' *Minerva* 1, 1962, pp. 54–73.

Rapoport, A. 'A Scientific Approach to Ethics' *Science* 125, 1957, pp. 796–99.

Ravetz, J. R. *Scientific Knowledge and its Social Problems* (Oxford, University Press, 1971).

Regan, T. & Singer, P. (eds) *Animal Rights and Human Obligations* (Englewood Cliffs, NJ, Prentice-Hall, 1976).

Rose, S. & Rose, H. 'Can Science be Neutral?' *Perspectives in Biology and Medicine* Summer 1973, pp. 605–24.

Schurr, G. *Science and Ethics* (University of Manchester, Science in a Social Context Trial Unit, 1977).

Smyth, D. H. *Alternatives to Animal Experiments* (London, Scolar Press, 1978).

Waddington, C. H. *The Ethical Animal* (London, Allen & Unwin, 1960).

Ward, B. & Dubos, R. *Only One Earth* (London, Penguin, 1976).

Weinberg, A. M. 'Criteria for Scientific Choice' *Minerva* 1, 1963, pp. 159–71.

Chapter 8 Science, Culture and Religion

Ashby, E. 'Universities Today and Tomorrow' *Listener* 65, 1961, pp. 959–60.

Barbour, I. G. (ed.) *Science and Religion: New Perspectives on the Dialogue* (London, SCM, 1968).

Cole, J. R. *Fair Science: Women in the Scientific Community* (New York, Free Press, 1979).

Goodman, D. C. (ed.) *Science and Religious Belief: A Selection of Primary Sources* (Bristol, Wright, 1973).

Hessen, B. 'The Social and Economic Roots of Newton's *Principia*' in N. Bukharin (ed.) *Science at the Crossroads* (London, Cass, 1971), pp. 146–212.

Hooykas, R. *Religion and the Rise of Modern Science* (Edinburgh, Scottish Academic Press, 1972).

Huxley, A. *Brave New World* (London, Chatto & Windus, 1932).

Isaacs, A. *The Survival of God in the Scientific Age* (London, Penguin, 1966).

Kearney, H. F. *Science and Change, 1500–1700* (London, World University Library, 1971).

Leavis, F. R. *Two Cultures? The Significance of C. P. Snow* (London, Chatto & Windus, 1962).

Marcuse, H. *One-Dimensional Man* (London, Routledge & Kegan Paul, 1964).

Orwell, G. *Nineteen Eighty-four* (London, Secker & Warburg, 1940).

Polanyi, M. *Personal Knowledge* (London, Routledge & Kegan Paul, 1958).

Ravetz, J. R. 'Criticisms of Science' in Spiegel-Rosing, I. & Price, D de Solla (eds) *Science, Technology and Society* (London, Sage, 1977), pp. 71–89.

Rose, H. & Rose, S. *The Political Economy of Science* (London, Macmillan, 1976).

Rose, H. & Rose, S. *The Radicalisation of Science* (London, Macmillan, 1976).

Roszak, T. *The Making of a Counter-Culture* (London, Faber, 1970).

Russell, C. A. (ed.) *Science and Religious Belief: A Selection of Recent Historical Studies* (London, University Press, 1973).

Skolimowski, H. *Eco-Philosophy* (London, Boyars, 1981).

Snow, C. P. *The Two Cultures and a Second Look* (New York, Mentor, 1964).

Steneck, N. H. (ed.) *Science and Society: Past, Present and Future* (Ann Arbor, Michigan, 1975).

Teilhard de Chardin, P. *Man's Place in Nature* (London, Collins, 1966).

Toulmin, S. 'The Historical Background of the Anti-Science Movement' in Wolstenholme, G. E. W. & O'Connor, M. (eds) *Civilisation and Science in Conflict or Collaboration?* (Ciba Foundation Symposium, Amsterdam, Elsevier, 1972).

Whitehead, A. N. *The Concept of Nature* (Cambridge, University Press, 1926).

Zuckerman, H. & Cole, J. R. 'Women in American Science' *Minerva* 13 (1975), pp. 82–102.

Chapter 9   The New Sociology of Science

Barnes, B. & Edge, D. *Science in Context. Readings in the Sociology of Science* (Milton Keynes, Open University Press, 1982).

Bloor, D. *Knowledge and Social Imagery* (London, Routledge & Kegan Paul, 1976).

Broad, W. & Wade, N. *Betrayers of the Truth* (New York, Simon & Schuster, 1982).

Butterfield, H. *The Origins of Modern Science, 1300–1800* (London, Bell, 1949).

Collins, H. M. *Changing Order. Replication and Induction in Scientific Practice* (London, Sage Publications, 1985).

Collins, H. M. 'A New Perspective for Science' *Times Higher Education Supplement*, 9 August 1985, p. 14.

Collins, H. M. & Shapin, S. 'Uncovering the Nature of Science' *Times Higher Education Supplement*, 27 July 1984, p. 13.

Cushing, J. T., Delaney, C. E. & Gutting, G. (eds) *Science and Reality. Recent Work in Philosophy of Science* (Notre Dame, Il, University Press, 1984).

Dean, J. 'Controversy over Classification: A Case Study from the History of Botany', in Barnes, B. & Shapin, S. (eds) *Natural Order. Historical Studies of Scientific Culture* (London, Sage Publications, 1979).

Fleck, L. *Genesis and Development of a Scientific Fact* (Chicago, University Press, 1979; first published in German in 1935).

Gale, G. 'Science and the Philosophers' *Nature* 312, 1984, pp. 491–95.

Glashow, S. L. & Georgi, H. 'Unity of Elementary-Particle Forces' *Physical Review Letters* 32, 1974, p. 438.

Hall, A. R. *The Scientific Revolution, 1500–1800* (London, Longmans, 1954).

Hanen, M. P., Osler, M. J. & Weyant, R. G. (eds) *Science, Pseudo-Science and Society* (Waterloo, Ontario, Wilfrid Laurier University Press, 1980).

Holton, G. 'Do Scientists Need a Philosophy?' *Times Literary Supplement*, 2 November 1984, pp. 1231–34.

Knorr-Cetina, K. D. & Mulkay, M. (eds) *Science Observed. Perspectives in the Social Study of Science* (London, Sage Publications, 1983).

Kohn, A. *False Prophets. Fraud and Error in Science and Medicine* (Oxford, Basil Blackwell, 1986).

Latour, B. & Woolgar, S. *Laboratory Life. The Social Construction of Scientific Facts* (Beverly Hills, Sage Publications, 1979).

Mendelsohn, E., Weingart, P. & Whitley, R. (eds) *The Social Production of Scientific Knowledge. Sociology of the Sciences*, vol. 1 (Dordrecht, Reidel Publishing Company, 1977).

Merton, R. *Social Theory and Social Structure* (New York, Free Press, 1957).

Mulkay, M. *Science and the Sociology of Knowledge* (London, Allen & Unwin, 1979).

Porter, R. 'White Coats in Vogue' *Times Higher Education Supplement*, 5 November 1982, pp. 12–13.

Putnam, H. 'Philosophers and Human Understanding' in Heath, A. F. (ed.) *Scientific Explanation* (Oxford, University Press, 1981).

Quine, W. V. *From a Logical Point of View* (Cambridge, Ma, Harvard University Press, 1953).

Shapin, S. 'History of Science and its Sociological Reconstructions' *History of Science* 20, 1982, pp. 157–211.

Whitley, R. *The Intellectual and Social Organization of Science* (Oxford, Clarendon Press, 1984).

Index